Final Theory of Everything

The Astonishing Universe

Russell Eaton

This is a paperback edition available internationally.

For more information, or to book an event,
please contact the publisher:
Email: mailto@deliveredonline.com
Website: www.deliveredonline.com

ISBN – Paperback: 978-1-903339-76-3

ISBN – Ebook: 978-1-903339-73-2

First Paperback Edition: March 2024

CONTENTS

*

Preface

We must always endeavour to eradicate bigotry and prejudice from science, and we must always be on guard when such pernicious influences come knocking at the door

Russell Eaton, author

Author's Apology

The author apologies for any hurt feelings or dismay caused by the revelations in this book. It is understandable that talk of Einstenian relativity being spurious may upset some people in the scientific community whose career and credibility depend on the veracity of relativity. No disrespect is intended by the author.

Introduction

This book reveals for the first time a new 'theory of everything' to explain how all aspects of the Universe are linked together, and why the Universe is the way it is.

The holy grail of cosmologists has been to find a master theory that provides a singular, all-encompassing, coherent theoretical framework of physics that fully explains and links together all aspects of the Universe.

So finding a theory of everything is one of the major unsolved problems in physics today. For the past two centuries cosmologists have been trying to find a common ground between the gravity we are familiar with and the gravity of subatomic particles. Now at last, that common ground can be revealed.

Some of the biggest unsolved mysteries in cosmology are now solved in this *Final Theory Of Everything:*

* Why dark energy and dark matter are non-existent and unnecessary in the Universe.

* Why there is no arrow of time.

* What keeps galaxies together.

* The ultimate fate of the Universe, and what happens before the Big Bang.

* The fictitious wave/particle duality of light.

* What happens inside a black hole.

* What lies beyond the standard model in cosmology.

* The fundamental nature of gravity that is common to big things and subatomic particles.

* A fifth fundamental force of nature that brings together the four known forces.

These and other mysteries of the Universe are unravelled, and the astonishing Universe that we inhabit is revealed for what it really is.

For example, ever since Isaac Newton published his Law of Gravitation in 1687 scientists have been trying to understand the underlying nature of gravity, but to date without success. We can measure gravity and feel it in our everyday lives, but what actually causes objects to gravitate towards each other? Newton could not explain it. Einstein tried to explain gravity by saying that when objects are close enough their mass disturbs an all-pervading ether, and that this disturbance makes such objects fall towards each other - but contemporary science dismisses the existence of an all-pervading ether. These pages reveal the actual underlying nature of gravity for both big objects and subatomic particles.

You now have the key to understanding the Universe like never before. A prediction is made that will revolutionise the exploration of the cosmos by revealing how we can put virtual video cameras on planets and stars however far away they may be (and instantly see the full video results).

This book is written for a general audience, for students, and in particular for astrophysicists and cosmologists who will gain new invaluable insights into the nature of our Universe. Be prepared for a shocking and exuberant rollercoaster ride through the Universe that busts many myths that we take for granted.

The Nature of Time

This section shows that time is nothing more than the measurement of movement. That there is no absolute phenomenon that is keeping time for the Universe.

When we use the word 'time' in our everyday lives, we are simply referring to movement (i.e. motion). To be more precise, time is the measurement of movement from A to B in relation to the movement of planet Earth (or some other frame of reference). Here are some examples:

- An hour of time is the measurement of movement of planet Earth moving on its axis one twenty-fourth of a complete rotation.

- A year of time is the measurement of movement of planet Earth moving around the sun to complete one full elliptical trip.

- A second of time as measured by an atomic clock is the measurement of the oscillating movement of an atom, and scientists have agreed that 9,192,631,770 such oscillations of movement equate to one second.

- A minute of time as shown on your mobile phone is the measurement of movement of one second, multiplied by 60. Normally, mobile phones receive their time from atomic clocks that make up the world-wide GPS system.

- A half hour of time as measured by a sun-dial clock is the measurement of movement of the sun's shadow travelling a certain distance around the face of a sun-dial located on Earth.

Note that these examples describe a measurement of movement, let's say from A to B for simplification. The key difference between 'distance' and 'time' is that distance is a measurement of movement, but only from A to B. Whereas time is a measurement of movement from A to B, but in relation to a given yardstick (to a given frame of reference with uniform periodicity). We humans have mostly decided to make that yardstick the movement of Earth.

To be more precise, time is just a concept; it is not a real thing nor an absolute phenomenon or manifestation. Put another way, time is a measurement of movement compared to another yardstick of movement, such as the movement of Earth. But that yardstick, i.e. frame of reference, can be something else if we so chose. For example it can be the movement of some other celestial body instead of planet Earth going around the Sun, or it can be the oscillations of an atom.

Another way to think of time is that it is a change of movement. We humans measure such a change of movement with the help of celestial bodies and atomic oscillations because they give us changes of movement with reliable periodicity.

In today's world we need very accurate clocks for navigation, GPS and so on. So scientists have agreed to measure time in terms of radiation emitted from atoms (i.e. oscillations) of the element cesium under specified conditions.

"The minute was formerly defined as the 60th part of an hour, or the 1,440th part (60 × 24 [hours] = 1,440) of a mean solar day—i.e., of the average period of rotation of the Earth relative to the Sun. The minute of sidereal time (time measured by the stars rather than by the Sun) was a fraction of a second shorter than the mean solar minute. The minute of atomic time is very nearly equal to the mean solar minute in duration". Source: Encyclopaedia Britannica.

So as mentioned, atomic clocks use a yardstick of movement to measure time. That yardstick (frame of reference) is the movement of certain celestial bodies. When the oscillations of an atom in such a clock are measured, the measurement of the movement starts at point A (the beginning of oscillation number 1) and finishes at point B (the end of oscillation number 9,192,631,770). That measurement of movements from A to B is then equated to a given movement of a celestial periodicity that equates to a second.

The following image shows how atomic oscillations (cycles) are measured per second in atomic clocks:

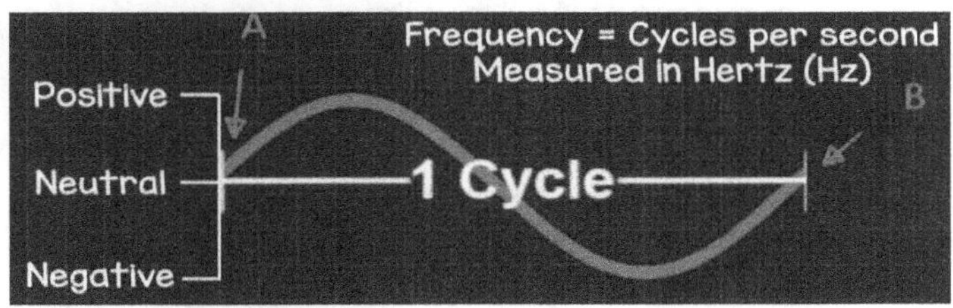

The word 'time' then is a human shorthand for describing *'the measurement of movement from A to B in relation to Earth's movement (or star/atomic movement)'*. A bit of a mouthful, so for convenience we just say 'time'.

There is no absolute phenomenon in the Universe that is responsible for the existence of time. There is no 'arrow of time'. The past and the future don't exist in any real sense, only the 'now' exists. There is no mysterious force or manifestation out there somewhere that is keeping time for the whole Universe.

We humans only know of three spatial dimensions in the Universe: length, width and depth. There may be more dimensions yet to be discovered, but time is certainly not a fourth dimension.

What about the ravages of time? Isn't time responsible for the degradation of our body as we grow older? If time does not make us grow old, then what does? The answer is movement.

Every part of our body is moving continually, non-stop. The molecules, atoms and subatomic particles that make up the human body never stop moving. In our daily lives we move about, eat different things and do different things. This changes the movement of some of the particles that make up our body. It is these minute changes in the movement of atoms, molecules, etc. inside our bodies that lead to ageing, degradation, illness, growth, good health, well-being, and so on. The so-called 'ravages of time' are entirely caused by this internal body movement and external events affecting such internal movements.

In passing it should be noted that the atoms in our body (and in fact atoms anywhere) never age in the sense of gradual degradation. There is no clock inside an atom telling it that it is now a minute older. Atoms do not experience time as an absolute phenomenon, but they do experience movement. Events can change the movement of atomic particles and in so doing the atom can disintegrate or change into something else. In this sense it can be said that atoms can decay through 'random' changes of movement, but not as a result of the atom becoming old or changing through time.

To clarify further, the movement (and changes in movement) in the human body cause all the changes that we experience, such as growing old, becoming weak or strong, ill or healthy, and so on. These physical changes in movement also affect human emotions; the way we feel, think and live from day to day.

When we say that a person has grown old in the last 5 years, what we are really saying is that a person has grown old during the last 5 rotations of planet Earth around the sun, and that during those 5 rotations, the person's body suffered (or enjoyed) many internal changes in particle movements.

We should realise that everything in the Universe is constantly on the move. Nothing is ever still. Everything in the Universe is continually moving in one way or another. A rock in the middle of the desert moves along with Earth's movements, and the atoms that make up the rock are moving continually. We humans measure certain movements, such as the movement of planet Earth around the Sun (or the movement of stars in relation to Earth), and we call such measurements 'time'.

The take home message: Time is not an absolute phenomenon or some kind of independent agent that is keeping time for the Universe. Time is a shorthand way of referring to the measurement of movement in relation to a given event of periodicity such as the movement of Earth or the oscillating movement of an atom.

The Calculation of time

This section shows the real nature of time when it comes to the mathematics of time. If you are not mathematically inclined, feel free to skip this section or just skim through it.

The formula for time

A formula for calculating time is readily available: Time = Distance ÷ Speed.

The Merriam-Webster dictionary defines 'time' as *'the measurable period during which an action, process, or condition exists or continues'*.

In the context of mathematics, to calculate time let's take the example of a car that moves one mile, from A to B. Here we are measuring a movement of one mile. If such a movement or motion takes two minutes, we say that the car takes two minutes of time to move one mile.

But it is equally valid to say *"The Earth moved 33.33 miles on its axis as the car moved one mile. And 33.33 miles of the planet's rotation is what we humans refer to as two minutes."* **Note:** The Earth rotates at about 1000 miles per hour on its axis. We call the 1000 miles of rotating movement 'one hour'. So 33.33 miles of rotation equates to two minutes. In our example then, the car moved one mile and during that journey planet Earth moved 33.33 miles. The two minutes of car journey *'time'* is simply a shorthand reference to the *'33.33 miles of Earth's rotation'*.

The formula for distance

What about 'distance' as given in the equation Time = Distance ÷ Speed?

The Merriam-Webster dictionary defines 'distance' as *'an extent of space'*.

In the context of mathematics, to calculate distance we use the formula: Distance = Speed x Time.

Distance is the measurement of movement of an object without any regard to direction. Thus distance can be measured along a curved or straight line. We said before that time is the measurement of movement from A to B in relation to Earth's movement. So to clarify:

Distance = measurement of a movement from A to B. Time = measurement of a movement from A to B in relation to Earth's movement.

Note: in this book we mostly use the word 'movement' rather than 'motion' but they both mean the same thing.

The formula for speed

What about speed as given in the equation Time = Distance ÷ Speed?

The Merriam-Webster dictionary defines 'speed' as: *"The rate (magnitude) of motion"*.

In the context of mathematics, to calculate speed we use the formula: Speed = Distance ÷Time.

Mere movement is simply a change in position of an object caused by a force acting on the object. Speed is the actual measurement of a movement. Put simply, movement is a change in position. Speed is a measurement of that change in position. So now we can summarise the three elements of the formula for time:

Time = The measurement of any given movement from A to

B, but in relation to Earth's movement (or some other yardstick such as the movement of stars or the oscillations of an atom). Example: When the Earth moves 33.33 miles on its axis, we call that amount of movement two minutes.

Distance = The measurement of any given movement that an object travels from A to B. Example: One mile.

Speed = The measurement of any given movement that an object travels in a given time. Example: One mile in two minutes. Or put another way: While the car moved one mile the Earth moved 33.33 miles on its axis.

We need to be careful to not confuse time with speed. Time is just a word that refers to gauging the movement of, say, planet Earth. Speed is a word that refers to a movement that is measured with such time, i.e. that is gauged with the movement of atomic oscillations or the movement of Earth.

As mentioned, the formula for time is: Time = Distance ÷ Speed. To calculate the time it takes a car to travel one mile at a speed of *one mile per 2 minutes*, we can write the formula as follows:

Time = One mile (distance) ÷ 2 minutes per mile (speed). Thus, if you drive your car at a speed of 1 mile per 2 minutes, and you travel a total distance of 1 mile, then you travel for 2/1 = 2 minutes.

You can equally say that if you drive your car at a speed of 1 mile per 33.33 miles of Earth rotational movement, and you travel a total distance of 1 mile, then you travelled for 33.33/1 = 33.33 miles of Earth-rotational-movement.

Or instead of saying you travelled for 33.33 miles of Earth-rotational-movement, you could say you travelled for X number of atomic oscillations (9,192,631,770 oscillations multiplied by 120 seconds). This of course is totally impractical, so for convenience and simplicity, we just use the word time and refer to 'two minutes'.

The time conundrum

In physics and cosmology there are many examples of equations that incorporate 't' (time) in their calculations, such as the following:

$$N(t) = N_0 \left(\frac{1}{2}\right)^{\frac{t}{t_{\frac{1}{2}}}} \rightarrow t = \frac{t_{\frac{1}{2}} \ln(\frac{N_t}{N_0})}{-ln2}$$

$$t = \frac{5730\ln(\frac{25}{100})}{-0.693} = 11460$$

In the above image it has been calculated that a fossil is 11,460 years old, that t = 11460.

In saying that 'time does not exist' a conundrum that arises goes like this: If time does not exist as an absolute phenomenon, what about the many equations that incorporate 't' (time) in their calculations? If 't' is omitted from such equations (by virtue of not existing), that would render the equations meaningless. That is the conundrum.

The answer to the time conundrum

The conundrum is resolved quite simply: In saying that time 'does not exist' what we are really saying is that time or 't' is nothing more than a measurement of movement. So all the equations involving 't' remain mathematically valid, and 't' for time can continue to be used without hindrance. If we take the above-mentioned example, we see that 't' is calculated as 11,460 years of time. This is the same as referring to a measurement of movement of Earth going round the Sun 11,460 times.

Put another way, the fossil has aged while Earth went around the Sun 11,460 times. Time as an absolute independent agent does not exist. But of course movement exists, and in our example, while the fossil lay in the ground during Earth's journey around the Sun 11,460 times, the *movement* of atoms in and around the fossil caused it to age.

Here is another example (from the many) of an equation incorporating 't' as time:

$$f(t) = a \times e^{kt}$$

This formula shows how to calculate exponential growth rates. For example, a scientist is studying the growth of a new species of bacteria. He wants to know how long it takes for the bacteria to grow to, say, a population of 500. Using the formula he works out that 't' is equal to 10 hours. This means that the bacteria multiplied to a quantity of 500 during a time of 10 hours. Since the Earth moves 1,000 miles on its axis in one hour, instead of saying 'the bacteria multiplied to 500 in 10 hours' we could say 'the bacteria multiplied to 500 while the Earth moved 10,000 miles on its axis'.

And here is a third example of using 't' for time in equations:

$$t = 1/H_0 = 1 / 2.37\,x\,10^{-18}\,1/s = 4.22\,x\,10^{17}\,s = 13.4\,billion\,years$$

This is an equation for calculating the age of the Universe. It shows that 't' is equal to 13.4 billion light years of time. In other words it took light from the Big Bang 13.4 billion years to reach Earth. This is the same as saying that while light travelled from the Big Bang to Earth, the planet Earth travelled around the Sun 13.4 billion times. But of course the Earth and Sun did not exist for most of light's journey from the Big Bang to Earth. It is simply that the 13.4 billion years of light-travel-time represents the equivalent of 13.4 billion journeys of the Earth around the Sun.

So yes, we can absolutely continue to use ‘t’ for time in mathematics in the same way that we have always done so. We simply regard ‘t’ or ‘time’ as a convenient shorthand for referring to a measurement of movement instead of a measurement of time as an absolute phenomenon.

In physics, time is defined by its measurement using a clock. And whatever type of clock may be used for such measurement, the clock will be measuring movement, whether it be a sun-dial clock, an atomic clock, or the moving hands of a grandfather clock.

The take home message: time is nothing more than the measurement of movement in relation to the periodicity of an event such as the periodicity of Earth’s movement or the periodicity of atomic oscillations. There is no past or future, there is no arrow of time, only the now exists.

The Now Theory

In what follows '*The Now Theory*' is revealed for the first time as a part of the jigsaw puzzle that makes up the *Final Theory of Everything*. There are two parts to *The Now Theory*. Part one is titled 'only moving things'. Part two is titled 'only the now'.

Part one: only moving things

We have said that there is no arrow of time going through the Universe, and the Universe is not endowed with some kind of universal time-keeping phenomenon.

The Universe does not age, evolve or change over time. Rather, the Universe ages, evolves and changes as a result of movement only. Everything that exists takes the form of 'moving things'. This is so without exception. For example, radiation, light, sound and human thoughts are all examples of moving things. When you think of something, you trigger certain neurons and molecules in the brain to form a thought. So a thought is a movement of neurons, molecules and chemical reactions.

Emotions are other examples of moving things. If you feel an emotion such as love, anger, tiredness, exuberance and so on, such emotions emanate from things moving inside the body. Physical things such as chemical reactions, electrical pulses, brain neurons and other moving things cause feelings and emotions.

Even concepts and memories are physical moving things. A concept or a memory is a thought or idea. And whatever you think of (call it concept, idea, imagination, creativity, memory, etc) it emanates from (it is based on) moving things inside the body. The whole Universe without exception consists of just moving things, nothing else. In this book you will discover a new theory of gravity for everything, from the big to the smallest subatomic particles. As you will discover, such gravity is caused entirely by movement.

As we have said, time is simply movement. To gauge a specific amount of time you do it by comparing a given movement (such as your daily commute to work) against an agreed event of periodicity such as the periodicity of Earth's movement on its axis and around the sun. You time your daily commute as 50 minutes, and of course the 50 minutes is another way of saying the Earth rotated on its axis 1,385 kilometres relative to the Sun. So time is nothing more than a moving thing, albeit in relation to an event of periodicity. And of course the periodicity event itself is a moving thing.

There is nothing in the Universe that is not moving. All of existence is entirely made of moving things, and only moving things. When new atoms or elements are created or changed it is simply part of the universal dance of moving things.

It is speculated that we live in a null-energy Universe (also known as a zero-energy Universe) as some scientists have declared. A null energy Universe proposes that the total amount of energy in the Universe is exactly zero: its amount of positive energy in the form of matter & energy is exactly cancelled out by its negative energy in the form of gravity. Some of the reasons for this speculation are as follows:

1. It allows for the creation of a Universe from nothing, without needing to posit a violation of energy conservation.

2. It fits in with the FTOE (Final Theory Of Everything) as postulated in this book. As mentioned, everything that exists is always moving as a result of cosmic expansion. This movement creates gravity and kinetic energy. This kinetic

energy from gravity can be thought of as negative energy. And this negative energy is what balances out the positive energy so as to give us a null-energy Universe.

3. When the big bang occurred, two possible scenarios arise: a null-energy Universe or a Universe that would have X amount of energy 'from the word go'. How was the particular X amount of energy determined? Would it have been possible to 'put' all the energy that we see in the Universe today into the very first moment of the Big Bang?

4. It obviates the need for the existence of dark energy because a null-energy Universe has the required balance of energy to resolve the issues proclaimed by those who seek the existence of dark energy. Equally, no dark energy is required because cosmic expansion is fuelled by a null-energy Universe rather than by some theoretical, unproven, undetectable energy called 'dark energy'.

Part two: only the now

Going on from the theory that the entire Universe is made of moving things (and nothing else), how does this fit in with the concept of time? We have said that time is nothing more than the measurement of movement from A to B in relation to some other movement of periodicity. Examples of movements of periodicity include the movement of Earth around the Sun, the oscillations of an atom, the regular movement of a given celestial body, and so on.

So for example a day of time is the measurement of movement of 40,000 km (Earth going around once on its axis). When this movement is related to the periodicity of Earth revolving on its axis, we call one such revolving movement a day of time.

So time is just a shorthand way of referring to moving things. But what about the past and the future? We know *"the past existed"* otherwise the present wouldn't exist. And we know *"the future will exist"* because we see ourselves moving into the future all the time as we get on with our lives and age.

But in the aforementioned, the reality is that the past never existed, and the future will never exist, but we use phrases like 'the past' and 'the future' colloquially in everyday language.

To clarify further, right now the past and the future do not exist. We only live in the present - only the now. We live in a continuous procession of now-moments. So right now everything is real and the Universe is real and it fully exists. And the Universe exists entirely as moving things.

When it is said that the past does not exist right now, it means there is no past. When our parents lived in the past, they in fact lived only in the now, just as you and I are doing so right now. So right now the past does not exist at all, and there has never been a moment when it has been possible to say that the past exists.

When we look at old buildings and historical monuments we are not looking at the past, we are looking at their present state. The same goes for fossils and anything else that points to a past existence. The past did indeed exist but only as NOW moments. History is full of clues about the past, but such clues only point to now moments, not to any kind of existential past at all, only the now.

The same goes for the future. Right now the future does not exist and it will never exist. There will never be a moment when we can say that right now the future exists, or that right now I am in the future.

We can of course affect or change the future. For example, I may decide to do something different tomorrow compared to my usual routine. This will affect what happens in my future-of-tomorrow compared to what would have happened if I had not done something different. Human endeavours such as creativity, progress, inventions and so on are all based on some kind of planning and changing of future outcomes. But we can never exist in what we call 'the future'. We can only ever exist in the now.

As an example, let's suppose I plan to have a party on my next birthday, so I make plans accordingly. I can only ever make those plans in the now. So I live in a continuous procession of now-moments in which those birthday plans exist. As the day of my birthday party approaches, those plans will come to fruition because I have changed things in the now (and in the now only).

Semantically we may say that the future is real. A man with terminal cancer may say that he will die within a year. A farmer may say that he will reap the harvest next Summer. A housewife may say that tomorrow she will cook a special thanksgiving dinner. All these things come to bear. But the fact that we do things that turn out to be so in the future is an illusion. When we do things that turn out to be so in the future, in reality we do things that turn out to be so in the now. We can never be in the future. We can only exist in the now, and the same goes for an alien living on a planet on the other side of the Universe. You and the alien exist in the same 'now' of the Universe. And what does that 'now' consist of? The mentioned 'now' consists entirely of moving things, nothing else.

When we look at ancient star light that has taken many light years to reach Earth it is said colloquially that we are looking into the past. If the light took 100 years of travel time to reach

Earth, we are looking at the starlight as it was 100 years ago. This does not mean that we are in reality looking at the past or into the past. It means we are looking at the streams of starlight as they are today. The same goes for cosmological photographs of galaxies, star formations and the like.

All such photos are not photos of the past, they are photos of light or infrared rays as they are today, at the moment they reach our eyes or the telescope/camera. We can never see the past because the past does not exist. All we can see is the now. When we look at photos of the past, we are in fact looking at photographs (whether on paper or digital) as they are today, regardless of what the images may represent.

The past and future are illusions or mental constructs. Only the now exists. But of course illusions and constructs do exist as moving things inside our brain in the way that thoughts exist. But what the illusions and thoughts actually represent do not exist if they are just in the mind.

Right now the whole Universe exists and we are a part of it. And everything that exists without exception is entirely made of moving things and nothing else. Right now no past or future exists. Semantically we may say the past existed by looking at history, but that does not mean the past exists now in any real sense. Only the present exists. Old books, historical monuments, ancient tree-rings all exist, but they exist as they are today and they serve as clues to the past. Equally we can speculate about how the future will be, but any such future will never exist as a 'future', it will only ever exist in the now.

For the sake of completeness we briefly discuss a phenomenon known as *'entropy as an arrow of time'*. This refers to an abstract concept that views time as an absolute phenomenon that only *'moves forward'* from the past, into the present, and then into the future. Hence, it is colloquially referred to as the arrow of time.

The way this arrow of time affects or relates to the Universe and the world we live in is referred to as a thermodynamic arrow of time. Thermodynamics is a branch of physics which deals with order & disorder, energy and the work of a system. So the thermodynamic arrow of time is defined as a one-way direction of time in which entropy (disorder) tends to increase with time.

The reasoning is that entropy (disorder) increases in the Universe with the passage of time. This disorder has made it possible for galaxies, planets and life itself to exist. Such entropy requires the passage of time to flow from the past to the present and into the future. But it does not follow that a thermodynamic arrow of time as an absolute phenomenon is required for this to happen. Even though the past and future don't exist, and we only live in the now, this in no way invalidates the concept of increased entropy with time.

As the past and the future do not exist except as concepts, there can never be some kind of absolute thermodynamic arrow of time. But of course, we can talk about an arrow of time colloquially when referring to the past, present and future.

The take home message: Everything in the Universe that exists takes the form of moving things, nothing else exists. We can only live in the now. The past and the future don't exist, only the now exists in the form of moving things. If we say *'we only live in the now'*, this means *'we only live as moving things'*.

Spacetime

In this section it is shown that spacetime, as defined by Einsteinian relativity, is spurious and non-existent.

In the special theory of relativity, Albert Einstein said that time is relative, that the rate at which time passes depends on your frame of reference. In other words, the amount of time that transpires depends on the movement or non-movement of the observer. The word 'observer' is used as a frame of reference from which a set of objects or events are being measured or compared against.

To clarify further, Einsteinian relativity is not saying that the act of observing an object move has an effect on the time experienced by said object. It is clearly postulated that time passes differently for each observer (each frame of reference, each object) depending on their location and how fast they are moving or not moving. This is disingenuous to say the least. Let's break this down into the following two points.

1. *'Time passes differently for each observer (each frame of reference)'*: If we take time to be a measurement of movement from A to B in relation to, say, the movement of Earth, then clearly (and contrary to relativity) time passes exactly the same for everybody (and for every frame of reference), given the same way of measuring time for everybody.

2. *'Time passes differently depending on the observer's location and how fast they are moving or not moving'*: If we accept the Einsteinian concept of time as an absolute phenomenon, then the inevitable conclusion is that every object in the Universe, every living being, every planet, every grain of sand, is enjoying its own unique timeline. Why? Because everything in the Universe is always moving, and relativity stipulates that the movement of an object ***in itself*** slows down time or affects the time experienced by that

object.

The special theory of relativity says or implies that movement of any sort, fast or slow, affects the time of that moving object. Given that everything in the Universe is continually moving (nothing is stationary or 'at rest'), then everything must be enjoying a slowing down of time in its own unique way.

A stone in the middle of the Sahara Desert is moving along with Earth's movement, so time is slowing down for that stone by the fact that it is moving, and any comparison to another frame of reference is irrelevant.

According to relativity, if you were to put a clock on top of said stone, it would show a different time to the clock that you carry, assuming the time-keeping of the clocks could be measured accurately, i.e. relativity postulates that the rock would show its own unique time and you would show your own different time, regardless of whether you are observing or not observing the rock, and regardless of your distance from the rock.

So the special theory of relativity says that in our daily lives real time differences exist between individuals (frames of reference) but are too small to be noticeable, i.e. too small to be measured with current technology. But at very high speeds, time is said to slow down noticeably. In relativity there is no universal time that follows some kind of timekeeping that is the same for everybody.

To illustrate this, the famous 'twin paradox' thought experiment has been postulated in various versions. We imagine two twins, twin A and twin B. Twin A travels around the solar system at nearly the speed of light. Meanwhile twin B sits at home watching TV most of the time. Having travelled extensively at nearly the speed of light for, say, about a week, twin A then goes home and greets his twin brother B.

In this scenario, the special theory of relativity says that when the twins meet again, twin A will have aged less than twin B by virtue of travelling very fast compared to couch-potato twin B who hardly moved. Put another way, time is said to have slowed down for twin A compared to twin B.

Is this so? Einsteinian relativity says the answer is YES, twin A aged less. If you accept the nature of time as described in this book, the answer is NO, time did not slow down for twin A because time is not some kind of 'independent' absolute phenomenon that exists and that can slow down in some manner. As mentioned, time is merely a human shorthand for referring to the measurement of movement from A to B with regard to a given yardstick, such as the movement of Earth.

Granted, it is possible that by travelling at very high speeds the human body may experience some biological changes caused by the high speed itself, but this has nothing to do with 'time slowing down' or making a person younger than his peers on Earth.

To finish with the twin paradox experiment, there is no paradox. Both twins aged at the same rate, albeit that twin A might have been affected biologically (in a good or bad way) by travelling at such high speeds.

To reiterate this important point, time is just a concept, a word invented by humans to describe a given measurement of movement (such as an hour) in relation to Earth's daily dance or in relation to the movement of stars. Time is not something real that is out there somewhere waiting for humans to find and understand.

When Einstein was thinking about these things he desperately needed to understand and explain the nature of gravity. It is well documented that in 1912 Einstein was actively seeking a theory in which 'gravitation' as he called it could be explained as a geometric phenomenon so he looked around for help. At the urging of Tullio Levi-Civita, Einstein began exploring the use of general covariance (which is essentially the use of curvature tensors) to create a gravitational theory.

This eventually led to the well-known ‘general theory of relativity’ based on proposing a so-called 'curvature of space'. This theory says that gravity is not an invisible force that attracts objects to one another. Rather, that gravity is a curving or warping of space. The more massive an object, the more it warps curved space around itself.

Note: Many scientific papers have been published on the subject of ‘*general covariance*’, and today general covariance is widely regarded as being vacuous in the context of physics. Put simply, the poor scientific view of general covariance puts doubt on the veracity of general relativity, but Einsteinian relativity continues to be accepted in the scientific community in the absence of a credible alternative.

To make his theory of gravity work Einstein had to add a fourth dimension to his equations which he called ‘time’. So you have the three spatial dimensions of space that we are all familiar with plus a fourth dimension of time that nobody truly understands. This allowed Einstein to postulate that gravity is caused by a curvature of space in time - that space and time (called spacetime) curves space in a way that causes gravity. By doing this he was able to make his equations work mathematically.

According to Einstein, spacetime is like a stage that remains in place whether actors are treading its boards or not. Even if there were no stars or planets dancing around, spacetime would still be there. So for Einstein, spacetime was a real thing – without spacetime space curvature and gravity would not exist.

At this point you may be shouting: *"But what about Einstein's field equations proving the existence of gravity as a result of the curvature of space?".* Later, in the section 'The AE Force and Gravity' we discuss why Einstein's Field Equations are not reliable.

Consider the following. Einstein was very clear in postulating that as an object moves, the movement itself curves space around such an object resulting in gravity. For this to work there had to be some kind of physical interaction between the object and the curved space around it. Without some kind of physical interaction the space around the moving object would not curve because such space would not detect the presence of the moving object.

Einstein reasoned that if an all-pervading ether (sometimes spelt aether) existed it would provide an interaction between a moving mass and the space around it, thus making it possible for space to be triggered into being curved.

Einstein thought long and hard about his ether conundrum, and after much ambivalence stated that *"we may say that according to the general theory of relativity space is endowed with physical qualities; in this sense, therefore, there exists an ether. According to the general theory of relativity, space without ether is unthinkable ....".*

So having previously denied the existence of an ether for many years, Einstein changed his mind, and postulated the existence of an ether. This allowed him to publish his general theory of relativity.

But modern-day physics has never detected the existence of an ether, and most scientists dismiss the idea of an ether, putting further doubt on spacetime and the curvature of space.

Some physicists consider that spacetime is merely a

mathematical concept (not an absolute phenomenon as defined by Einstein) that is used to calculate the 'curvature of space' as proposed by relativity.

Here is a description of spacetime curvature, whether it be real or a mere mathematical concept:

"The curvature of spacetime influences the motion of massive bodies within it; in turn, as massive bodies move in spacetime, the curvature changes and the geometry of spacetime is in constant evolution. Gravity then provides a description of the dynamic interaction between matter and spacetime". Source: European Space Agency. **Note:** the curvature of space is said to make objects go towards each other (the 'force' of gravity).

If you search internet for a description of spacetime you get a variety of answers:

* Spacetime is a graphical illustration of the properties of space and time in the special theory of relativity.

* Spacetime is any mathematical model which fuses the three dimensions of space and the one dimension of time into a single four-dimensional manifold.

* Spacetime is a mathematical model that joins space and time into a single idea called a continuum. This four-dimensional continuum is known as Minkowski space.

* Spacetime explains the curvature of space, and this curvature of space is responsible for gravity.

Einstein's equations are said to show how spacetime makes space curve, thus causing gravity. So spacetime is said to explain this curvature of space. For example, if an asteroid or satellite comes too close to Earth it will fall into a so-called curvature of space that exists around Earth, and this curvature will make the asteroid or satellite fall towards Earth. The curvature of space caused by spacetime is often illustrated as a grid on a 'trampoline':

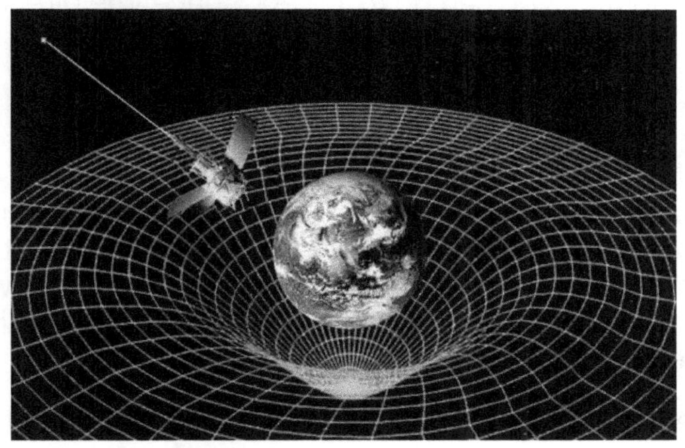

As mentioned, this concept of space curvature as postulated in the general theory of relativity can only 'work' mathematically if a fourth dimension called 'time' is added to the equations. This mysterious phenomenon called 'time' has never been shown to exist, and to this day cannot be explained at a fundamental level (because it is simply not understood). Whether spacetime is a mathematical concept or a real thing becomes irrelevant because if time does not exist as an absolute 'independent' phenomenon, then spacetime cannot exist in the absence of this phenomenon.

Here's a quote from Sir Karl Popper, one of the 20th century's most influential philosophers of science:

"A time dimension would make motion impossible. Nothing can move in spacetime for this reason. Spacetime is a non-existent block Universe, an abstract mathematical construct. We live in the continually changing present always, the Now. And no, we are not moving in time from the past toward the

future either".

It is increasingly being accepted by scientists in today's world that spacetime is not real in the same way that an atom is real. There's nothing you can do to 'detect' spacetime directly, they say.

Where many scientists go wrong is to go on to say that although spacetime may not be real, spacetime acts to curve space in a very real way, and it predicts things such as gravity, time dilation and the bending of light. This book explains why this is not so and provides a new theory of gravity that serves equally well for stars and galaxies as for the quantum world of subatomic particles. But first we will look at time dilation and then the *Majesty of Light*.

For the sake of accuracy, it should be mentioned that the concept of 'spacetime' is not contained in Einstein's 1905 paper on Special Relativity. Rather, it was 'invented' in 1908 by Hermann Minkowski, one of Einstein's mentors. Spacetime emerged from a mix of (1) Lorentz's ad hoc ether concepts, (2) the contraction of matter and the Lorentz transformations, (3) Einstein's ad hoc concepts of Special Relativity, and (4) Minkowski's imagination and manipulated mathematics.

Thus, Einstein very much based his concept of spacetime on Lorentz's contraction hypothesis, which in turn was based on the existence of a stationary ether. The fact that an ether does not exist means there can be no valid measurements from it. Quite simply, and to put it crudely, spacetime is fanciful nonsense.

For these reasons alone, spacetime is ad hoc, empirically invalid and meaningless as such. During the last century, the concept of spacetime geometry has served as an inspiration for physicists to explain, analyse and illustrate various theories of relativity. Unfortunately this has also served to hold back the advancement of cosmology and our understanding of the Universe.

The take home message: spacetime and the curvature of space are concepts that are baseless, meaningless and non-existent.

Time Dilation

This section shows that time dilation as postulated in the special theory of relativity is spurious and non-existent.

This is how Wikipedia refers to time dilation:

"In physics and relativity, time dilation is the difference in the elapsed time as measured by two clocks. It is either due to a relative velocity between them or to a difference in gravitational potential between their locations. When unspecified, time dilation usually refers to the effect due to velocity".

Time dilation is taken to mean that 'time slows down'. But this subject causes much confusion and head scratching. For example it is said that the faster the relative velocity, the greater the time dilation, with time slowing to a stop as one approaches the speed of light.

Those who profess time dilation say that *"time dilation means that the time between two events is different for observers moving at different speeds".*

What does this really mean? They seem to be saying, for example, that if John drives his car from point A to point B at a constant speed for 2 minutes, then John perceives that the journey took two minutes. But if Jill views the same car trip from a nearby hill top she would perceive that the journey took longer than two minutes.

The reasoning here is that time slowed down for John while driving his car by virtue of time dilation caused by the car movement itself. So for John the trip took a real two minutes. But for Jill atop her hill, sitting among her very accurate time measuring equipment, the trip took longer than two minutes, it took, say 2.5 minutes because she was not moving relative to John. **Note:** Jill measured *her* time, starting and finishing with the car movement.

In this example, John enjoyed time dilation by virtue of his high speed car, so for John the journey took two minutes. But the stationary observer (Jill) measured the car journey as taking, say, 2.5 minutes. Two different time measurements for exactly the same car journey.

Granted that time dilation becomes too negligible to measure at low speeds if at all, but even if John drove at 90% the speed of light for 2 minutes, it is absurd to think that the timed event that occurred for John changed intrinsically for Jill. Or vice-versa, that the event accurately timed by Jill is different from John's event. That we have two fundamentally different events emanating from the same car journey.

When physicists are confronted with examples like this you get comments such as: *"The event didn't change in itself, but John and Jill each had a different perception of the same event".* If this is the case then we are only talking about human perception and time itself did not slow down! But if time did not slow down, then that would mean that time dilation did not occur, that it is not a real absolute phenomenon.

It is also said in Einstein's general theory of relativity that gravity can bend spacetime, and therefore time itself. The closer the clock is to the source of gravity, the slower time passes; the farther away the clock is from gravity, the faster time will pass.

So which is it? Is time dilation caused by speed or is it caused by gravity, or by both? By now you will appreciate that if time does not exist as an absolute 'independent' phenomenon (or as a separate fourth dimension), then time dilation cannot be caused by Einsteinian gravity. Why not? Because if Einsteinian gravity cannot exist without a fourth dimension of time, then such gravity cannot be a cause of time dilation.

So we're left with speed as a cause of time dilation. Scientists have made great efforts to demonstrate time dilation in various ways, but when examined, a credible demonstration of time dilation is lacking, and alternative explanations are readily available.

In physics it is claimed by some that 'thousands of tests' show time dilation to be real and that they conform to Einsteinian relativity. In fact if you delve into this subject you don't find thousands of such tests, rather you find a few dozen claims published in scientific papers attempting to show that time dilation is real. If you then drill down into these claims you find that they don't demonstrate or 'prove' time dilation at all.

In most attempts to prove the veracity of time dilation, gravity is responsible for the experimental results rather than showing that they were caused by time dilation. In reality, experiments involving so-called 'gravitational time dilation' can readily be explained by the effect of gravity itself rather than the effect of time dilation.

Note: Gravity affects the speed of clocks. The stronger the gravity, the greater the so-called 'treacle effect'. As you move away from Earth, gravity becomes weaker and this makes any kind of time-keeping mechanism including atomic clocks run more freely (less treacle), and hence run faster than clocks at ground level. Also, air density/pressure decreases rapidly with increasing height and this can greatly make clocks run more freely and quickly.

When it comes to time dilation caused by speed (referred to in physics as kinetic time dilation), subatomic particles are used in experiments because of their high speed. But the few experiments conducted to prove kinetic time dilation have been unreliable or inconclusive.

For example, it is postulated that light contracts in physical length at light speed. But the contraction hypothesis as formulated by the Lorentz–FitzGerald contraction is regarded as false for the very fundamental reason that it entails a contradiction, namely, the consequence that light must have a variable velocity along what by definition is taken to be a rest length. Furthermore, the 'Lorentz Contraction' of matter hypothesis very much depends on the existence of an ether which we now know does not exist.

For the sake of completeness, what follows are some of the efforts made to demonstrate the existence of time dilation. But in this book we will not waste time (excuse the pun) to analyse or disprove, one by one, every relativistic claim by Einstein and his disciples. The reader should not accept relativistic claims of any sort on face value.

Experimental Storage Ring vis-a vis time dilation

In September 2016, a paper published in 'Physical Review Letters' prompted worldwide headlines to the effect that the veracity of time dilation had been proved. An international group of collaborators including Nobel laureate Theodor Hänsch, director of the Max Planck optics institute, had worked for many years to test the time dilation effect.

They tested two clocks, one that is stationary and one that moves. To do this, the researchers used the Experimental Storage Ring, where high-speed particles are stored and studied at the GSI Helmholtz Centre for heavy-ion research in Darmstadt, Germany.

The scientists made the moving clock by accelerating lithium ions to one-third the speed of light. Then they measured a set of transitions within the lithium as electrons hopped between various energy levels. The frequency of the transitions served as the 'ticking' of the clock. Transitions within lithium ions that were not moving served as the stationary clock.

A very small time-difference between the two clocks was observed, but the scientists readily admitted they needed a larger accelerator so as to do more powerful tests. It could not be shown whether the results were or were not within a margin of error, and the experiment was discontinued.

Global Positioning System vis-a vis time dilation

The GPS (Global Positioning System) is described as a satellite-based radionavigation system owned by the United States government and operated by the United States Space Force. The GPS is a group of 31 atomic clocks put into orbit in a way that synchronises their highly accurate time-keeping with clocks on Earth.

It is widely believed that GPS depends on time dilation to be able to maintain this mentioned synchronisation. The belief, albeit incorrect, goes something like this:

"GPS confirms the existence of time dilation by allowing the timing and position of signals from moving orbital satellites to be synchronised with cell phones and electronic devices on Earth's surface, thereby providing accuracy for many activities on Earth such as car navigation".

Put simply, GPS does not employ or depend on any kind of time dilation. Here is an explanation:

All atoms are affected by gravity - this is universally accepted and undisputed, otherwise atoms would not stick together and there would be no stars and planets. This means the atoms being measured by atomic clocks are affected by gravity. As you move away from Earth, gravity becomes weaker.

To emphasise this important point, atomic clocks (and in fact mechanical or electronic clocks) run faster at high altitude because a weaker gravity allows the ticking or measuring mechanism a freer, less constrained movement.

In fact, the GPS atomic clocks, operating at 20,000 km above sea level, have a gravity that is about 4 times weaker than gravity on Earth. A weaker gravity makes atomic oscillations run faster to the effect that satellite clocks run about 45 microseconds faster than on the ground. Engineers allow for this in setting the time of the atomic clocks, and are therefore able to ensure that GPS clocks are synchronised with the movement of Earth, and hence with time at ground level. In fine tuning the atomic clocks engineers also take into account the weaker air density and temperature at high altitude, as these factors can also affect the speed of time-keeping.

Another factor is the movement of the GPS satellites as they orbit Earth. Their high speed of 8,700 mph affects the atomic clocks and makes them run about 7 microseconds a day slower than clocks on Earth.

This happens because the GPS satellites always have to trilaterate (similar to triangulate) a point on Earth to determine where to aim the time signal. Note: trilateration measures distances, not angles.

The following image shows how three GPS satellites work together and measure distances to determine the nearest receiving station *location* on the ground for sending the time signal. While doing this they continue to orbit at 8,700 mph.

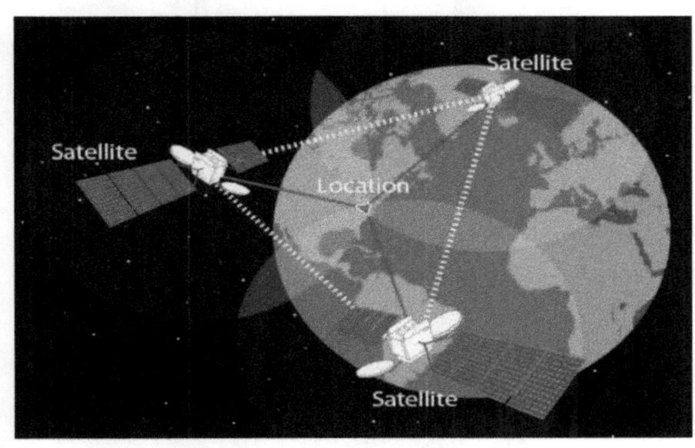

After calculating distances and trilaterating which ground receiving station will receive the time signal, the signal is then actually sent to that chosen ground station. But between calculating the signal's aiming point and then actually sending it, the satellites will have moved along in their orbit (they are not geostationary, they complete two full Earth orbits per day).

Thus, during the interval between finishing the trilateration and actually sending the time signal the satellites have moved along their orbit a certain distance from A to B. As a result of the satellites movement, the distance from point B to the Ground Receiving Station (GRS) on Earth is always going to be a longer distance than from point A to that same GRS. This longer distance creates a delay of 7 microseconds which has to be accounted for.

In the following image, the satellite has travelled from A to B. The distance from B to the GRS is longer than from A to the GRS:

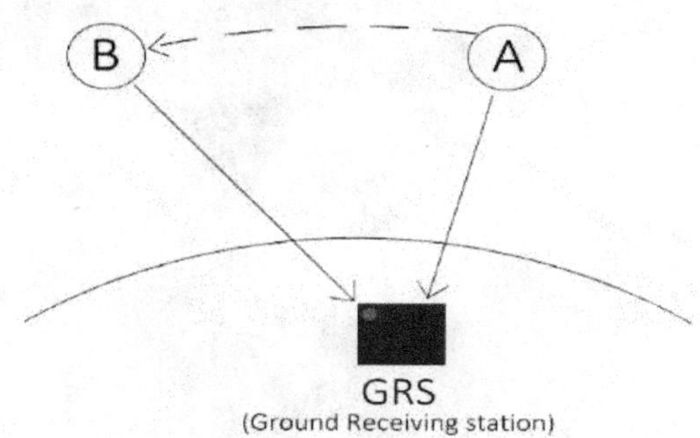

Normally, three GPS satellites do the trilateration and a fourth satellite looks after time adjustments to ensure good synchronisation between the atomic clocks and the time stamp received on Earth. Also at ground level there is GPS equipment to correct errors and ensure good synchronisation.

To summarise, the code inside the GPS satellites uses programs to ensure that timekeeping is synchronised with timekeeping at ground level. The satellites experience around 7 microseconds of 'delay' due to their orbital movement and about 45 microseconds of going too fast due to operating at lower gravity and the effects of atmospheric density and temperature. The coding (programming) of the atomic clocks compensates for these factors at all times. The body responsible for this coding readily admits that time dilation equations are not used for GPS:

"The Operational Control System (OCS) of the Global Positioning System (GPS) does not include the rigorous transformations between coordinate systems that Einstein's general theory of relativity would seem to require". Source: Henry F Fliegel, et al, GPS and relativity: an engineering overview, GPS joint program office, The Aerospace Corporation, USA.

Equally, the US military does not use time dilation in its GPS' calculations. Its scientists work with physics, not with imaginary phenomena. So-called time dilation has never had anything to do with GPS.

Regarding the Galileo Navigation Satellite System (GNSS), it also does not use time dilation equations in any of its systems. Reports that GNSS experiments have 'proven' time dilation with great accuracy refer entirely to gravitational time dilation (not kinetic time dilation). As already discussed, gravitational time dilation can be entirely accounted for by the slowing-down effects of gravity (the treacle effect) on time-keeping apparatus, rather than any kind of slowing down of time. Also, air density/pressure decreases rapidly with increasing height and this can greatly make clocks run more freely and quickly.

But even if special relativity and time dilation were to be applied to the GPS network it would face a major contradiction as follows. We have moving atomic clocks in orbit and stationary clocks at ground level; both types of clocks must continually be synchronised for the GPS system to work.

Special relativity predicts that time dilation will make GPS (atomic) clocks in orbit tick more slowly than ground clocks because GPS clocks are moving fast in relation to ground clocks at rest. But special relativity predicts that ground clocks will also tick more slowly than GPS clocks because ground clocks are moving fast in relation to GPS clocks at rest.

Something can only be said to be moving if it is compared to something that is not moving, or moving more slowly. So it is quite valid to say that GPS satellites move fast in relation to the Earth 'at rest'. And it is equally valid to say that the Earth moves fast in relation to the GPS satellites 'at rest'.

Special relativity is very clear in saying that time dilation must always apply to both objects, the moving object and the object at rest. As mentioned, nothing can be said to be moving unless it is compared to a non-moving object. The claim that GPS clocks are able to be synchronised with ground clocks by allowing for time dilation is incorrect. In other words, it is incorrect to say that time dilation causes GPS clocks to slow down and that therefore a time correction can be built in.

It is incorrect because it would be equally valid to say that ground clocks tick more slowly by virtue of moving fast in relation to GPS satellites 'at rest'. To resolve this and similar conundrums, Einstein's special relativity has to say that both GPS clocks and ground clocks will be slowing down in relation to each other because both types of clocks are affected by time dilation, not just the GPS clocks. This is the major contradiction, because it means there is no scope for building in a time correction.

To finish on the subject of GPS, it is interesting to note that the inventor of atomic clocks (upon which GPS is based) was adamant that special relativity is spurious. His name was Louis Essen, inventor of the atomic clock and the man responsible for the modern precise measurement of the speed of light. Essen published a paper titled *"The special theory of relativity: A Critical Analysis'* in which he readily criticised the theory as being wrong and that it hinders the development of a rational extension of electromagnetic theory.

Clock comparison vis-a-vis time dilation

According to the special theory of relativity, we have mentioned that time is said to slow down with movement. This kind of time dilation is called 'kinetic time dilation' (we use the term 'kinetic' rather than 'kinematic' in this book). To prove this, clocks have been flown around the world and then compared to identical clocks that were not moved. The well documented Hafele–Keating experiment comes to mind. Similar experiments using atomic clocks have also been conducted.

The results of such experiments do indeed show time differences between the moving clocks and the stationary clocks. How can this be?

A plane must be built so that lift and thrust are stronger than the force of gravity and drag by just the right amount. Lift from the wings is used to overcome the force of gravity. Takeoff and landing greatly increase gravity inside the plane. Shape is important in overcoming drag. All these things mean that gravity, however momentary, is very much fluctuating on a plane compared to, for example, gravity on the ground.

So when you take a clock aboard an aircraft, the clock will experience gravity fluctuations (however little) compared to its twin-clock on the ground. A greater force of gravity slows down the mechanical, electronic and atomic ticks of clocks. Metaphorically, gravity is like treacle - it slows things down.

The simplest time dilation effect to test is so-called 'gravitational time dilation', the speeding up of clocks with altitude. According to Einstein, the rate of time follows the gravitational potential, so a clock on the surface of the earth should count more slowly than a clock at some high altitude, such as at the top of a mountain.

This was tested in the 'Pico del Teide Experiment' in 2010 and it was concluded that time dilation did not occur.

"To believe that time dilation is real would require us to assume that walking up the mountain is the same as walking into the future - the land would have to form a bridge between two points in Einstein's dimension of time. The entire premise that time dilation corresponds to a change in 'real' time, and

that General Relativity can explain the (time dilation) effect, is highly questionable. If time dilation is an illusion, then the entire 4D time-space continuum of Einstein becomes superfluous". Source: Doug Marett, The Sagnac Effect: Does it Contradict Relativity? 2012, conspiracyiflight.com.

Many experiments show unequivocally that at high altitude, such as in satellites and in space, clocks run faster by being less constrained by 'the treacle' of gravity. They run faster not because of 'gravitational time dilation' but because less gravity allows a clock to run with less constraints upon whatever mechanism it is using for timekeeping. Also, air density/pressure decreases rapidly with increasing height and this can greatly make clocks run more freely and quickly.

This is why clocks on the International Space Station (ISS), for example, run marginally faster than reference clocks back on Earth. Claims that ISS clocks run slower than clocks on Earth are uncorroborated.

In reality there is no accurate way to compare clocks in different locations and different states of motion. You can send radio signals back and forth to compare them, but depending on the orbital parameters, they may each 'see' the other as running slow.

To make an unambiguous clock comparison that does not depend on choosing some reference frame as the real & true one, you would need to start with the clocks together and synchronised and then after you ride on the ISS for a while, bring them back together and compare. In that case it would be found that the ISS clock runs faster, not slower, by virtue of having less gravity.

But even doing this kind of experiment would be very difficult, because just the act of separating the ISS clock from the ground clock would introduce time distortions due to gravity, physical movement, environmental temperature, etc. ISS clocks do not employ any kind of time dilation factors in their timekeeping, they simply synchronise their time with GPS.

Claims that the amount of time difference between the moving and stationary clocks (in the mentioned experiments) was in accordance with relativistic predictions should be taken with a big pinch of salt when you take into account the margin of error and factors such as aircraft cabin pressure, vibrations, changes in temperature, gravity fluctuations and so on. All these factors can affect timekeeping apparatus. And furthermore, we must remember that special relativity predicts that both types of clocks (aeroplane clocks and ground clocks) will be equally affected by time dilation, and hence any comparison between the two types of clocks is meaningless.

Because of these experimental uncertainties, no further similar experiments have been conducted since the 1970's.

To summarise, there is a growing consensus among scientists that both time dilation and length contraction are invalid concepts because they are based on a set of equations that are inconsistent with physical principles. This inconsistency is also evident in the well-known 'Twin Paradox': speed is relative to something that is not moving, but time dilation also affects speed so there is a conflict.

Muons vis-a-vis time dilation

It is argued that time dilation can be demonstrated by the way subatomic particles called muons fall to Earth. This is how Wikipedia talks about muons in the context of time dilation (abridged extract):

"When a cosmic ray proton impacts the upper atmosphere, muons are created and they go towards Earth at nearly light

speed. Without time dilation the muons would have a half-life of about 456 metres. This means that every 456 metres the number of muons travelling to Earth is reduced by half (because half of them die), so very few muons would end up reaching Earth. But time dilation allows the muons to have more time to reach Earth; their time slows down, giving them a longer half-life, thus allowing them to reach Earth before dying. From the muon's point of view its time has not changed (time has not slowed down), but for the muon the distance to Earth has shortened due to time dilation, allowing it to reach Earth alive. From the viewpoint of Earth, the muon's distance to Earth has not changed, but time dilation has allowed the muon to live longer, thus enabling it to reach Earth." Source: Muon, Wikipedia.org.

Note: Special Relativity says that time dilation has a dual effect: it slows down time and it physically contracts (shrinks) the length of the object enjoying time dilation. Special relativity does not say that the distance between two objects contracts. Hence, in terms of Special Relativity, Wikipedia is incorrect to say that time dilation reduces the distance between the muons and Earth's surface.

In a bid to keep things simple we will avoid mathematics as much as possible and simply say that large amounts of muons arrive at sea level, not because of time dilation, but for five main reasons:

1. Favourable atmospheric conditions. The abundance of muons reaching Earth is mostly due to favourable weather conditions, not time dilation. Many more muons will reach Earth on a cloudless sunny day compared to a rainy or stormy day.

"Cosmic rays are mostly high-energy protons and are constantly bombarding atoms in Earth's atmosphere to create

pions. These pions either decay into lighter muons or continue to interact with nearby atoms and avoid decaying into muons. If the atmosphere is cool and thick then the chance of continued interactions is much higher, and the number of muons generated is therefore far fewer than when the atmosphere is warmer" (source: Muons reveal upper atmosphere's temperature, Physics World).

2. Speed of light. Muons reach the Earth's surface for the simple reason that they travel fast enough to do so. They travel fast enough (99.99% the speed of light) to reach the planet before decaying. This is much faster than muon speeds achieved in particle accelerators (about 90% the speed of light).

There is a great variety of speeds among subatomic particles, depending on their mass, temperature and composition. The muons that hit Earth result from particles in the Earth's atmosphere colliding with cosmic rays—high-energy protons and atomic nuclei that move through space at just below the speed of light. This creates highly energetic muons, unlike the slower muons created in particle accelerators.

In fact the majority of cosmic ray muons decay during their rapid flight through the atmosphere, but the most energetic of them survive long enough to not only reach Earth, but to penetrate deep underground. On the other hand, muons artificially produced in the laboratory on Earth have a much shorter lifespan of 2.2 microseconds when they are 'at rest' and then they decay into another particle called a 'positron'.

3. half-life of muons. It is argued that the half-life of muons would kill them off before reaching Earth if it were not for time dilation. Half-life is the expected time when half the number of atoms (or muons) have decayed, on average. No doubt many muons die before reaching Earth, but the sheer quantity of muons ensures that many survive to reach Earth in spite of their 'half-life'.

Furthermore, the half-life of muons has only ever been measured in particle accelerators at 1.56 μs under very controlled conditions. The half-life of atmospheric muons is thought to vary widely. If we take the half-life for muons to be 2 microseconds (2 millionths of one second). This means that one half of all muons only exist this long. But the other half exist for varying periods, including twice as long, 4 times as long, 8 times as long, 16 times as long, etc., and thus they travel much farther than the average.

The arithmetical calculations of a muon's half-life are only rough estimates because atmospheric conditions determine the length of half-life. Furthermore, half-life also depends on randomness because the atoms (and muons) decay at a random time. In short, the half-life of many atmospheric muons is likely to be longer than the half-life of particle accelerator muons, thus allowing such muons to reach Earth.

It is therefore disingenuous, to say the least, to present a set of half-life equations and argue that they demonstrate that without time dilation, half-life would kill off nearly all muons before reaching Earth.

4. Atmospheric muons versus lab muons. The Standard Model of Particle Physics is scientists' current best theory to describe the most basic building blocks of the Universe. It explains how particles called quarks and leptons make up all known matter. It also explains how force-carrying particles, which belong to a broader group of bosons, influence the quarks and leptons (a muon is part of the lepton group).

The latest research into the nature of muons indicates a tiny difference between how muons should behave according to the Standard Model, and what they were actually doing inside the giant accelerator. The research also shows that when an elementary particle such as a muon travels with a very high velocity, its lifetime (and hence its half-life) increases. Source: Patrick J. Kiger, Muons - The Subatomic Particles Shaking Up the World of Physics, Fermilab, June 2021.

This research shows two things:

A. We cannot assume that laboratory accelerator muons behave the same way as atmospheric muons. Hence, equations showing that muons can only reach Earth through time dilation cannot be relied on.

B. The fact that very high velocity increases the lifetime of muons adds to the argument that muons travelling at just below the speed of light are quite capable of reaching Earth. It is entirely possible, indeed, very likely that a significant proportion of cosmic muons are able to travel a distance of about 16000 metres in their life span to reach Earth without the aid of time dilation.

5. The muon time dilation contradiction. Special relativity says that when a muon travels at near the speed of light the muon experiences a slower time rate than that of an observer (the 'observer' being the planet Earth frame of reference). Not only that, relativity says that a muon travelling at 99.99% the speed of light will experience that time just about comes to a stop.

But you cannot move unless you have time to move. That being the case, how is it that muons nevertheless have enough time (i.e. 30 microseconds) to reach earth?

Physicists say it takes a muon about 30 microseconds to reach Earth. But it is also said that the muon does not experience the passage of time, i.e. it does not experience the 30 microseconds of time in its journey to Earth because at the speed of light no time passes.

To try and resolve this contradiction some physicists will say that although time has virtually stopped for the muon, the distance between the muon and Earth's surface has contracted for the muon. That instead of seeing, say, 10 km down to Earth, the muon 'sees' only 1.4 km (a contraction of about 7 times). This allows the muon to reach Earth in practically no time, even though a human observer would see the muon reach Earth in 30 microseconds.

Note: Special Relativity does not say that the distance between two events contracts, it says that the length of the speeding object contracts. This would require the contraction of the length of the muon itself. And (logically) a shorter muon would not be more likely to reach Earth than otherwise.

Another aspect of the mentioned contradiction goes as follows:

Special relativity clearly says that time slows down for moving objects. So when a muon speeds towards Earth, special relativity says that time slows for both the muon and the approaching Earth. From the moon's perspective, Earth is speeding up towards it in relation to the muon that is at rest. And equally, from Earth's perspective, the muon is speeding down towards the ground in relation to the Earth that is at rest. So special relativity says that as both objects are moving (the muon and the Earth), they must both experience time slowing down.

So in the scenario of the muon being at rest in relation to the ground moving towards it (a scenario required by special relativity) time does not slow down for the muon because it is not moving. Yet it is argued that muons manage to reach earth (before decaying) because time has slowed for the muon, giving it more time to reach the ground before decaying. This is the contradiction. You cannot have a muon at rest (with less time to decay before reaching the ground) and at the same time, that same muon is not at rest (with more time to decay before reaching the ground).

There is only one reality to the Universe, whatever it may be. It cannot be that on the one hand we have a muon taking 30 microseconds to reach Earth, and on the other hand we have that same muon travelling at the speed of light but frozen in time. It is impossible for something to move unless it has the time to make the movement.

To say that there can be more than one reality for the same given event is fanciful - the fruit of human imagination and contrived mathematics. But of course, just about any given event can be perceived by humans in many different ways. There's no accounting for human imagination and perception!

Speed of light vis-a-vis time dilation

As mentioned, according to the special theory of relativity, the faster something travels, the slower time moves for it. So for example, if an astronaut leaves earth and travels close to the speed of light, he will feel like 2 years have passed on his journey, but when he returns to earth, 40 years will have passed. So time dilation caused by the movement of an object is said to be a real thing.

Furthermore, it is postulated that when you reach the speed of light, time has slowed down so much that it comes to a stop. But light needs time to move. If time stops, how can light continue to move? Light moves at 186,000 miles per second. So it needs a time of one second to travel a distance of 186,000 miles. But if time stops at the speed of light, where does that one second of time come from? This is how Einsteinian relativity attempts to answer this question:

We know that light moves because we can see it moving with our instruments, and we can readily measure the speed of light. That is why we know that light moves at about 300 million metres per second. But from the perspective of light (for example imagine you are a photon of light), relativity says that time does not pass - time does not exist for light. As a photon of light you are standing still among other photons also standing still.

Nothing can move unless it has the space to move and some time in which it can make the movement. The contradiction is that if time stops for light, then such light cannot move. And of course we know that light does indeed move, and that it does indeed have time to move because we have measured the time of light's movement.

Relativity advocates may say that light, which indeed moves at the speed of light, experiences no time because time is frozen. But Einsteinian relativity itself says this interpretation is wrong. Time is not frozen, it is simply that there is no valid reference frame with which to compare the speed of light (nothing but sophistry).

A reference frame that has exactly zero spatial width and exactly zero time elapsing is simply a reference frame that does not exist. If an entity is zero in every way we try to describe it, how can we possibly say that the entity exists in any meaningful way? We cannot.

Such arguments reflect the contradictions inherent in time dilation, and the ambivalence among physicists to accept time dilation.

Ironically, Einstein's second postulate that the speed of light is independent of the motion of its emitting body (for example the Earth) is correct. But Einstein also asserted that the motion of the Earth through space affects how light propagates, making light take longer to travel due to time dilation. So there is a glaring contradiction that violates Einstein's second postulate which states that the propagation of light emitted on the Earth's surface is independent of the Earth's motion.

The 'standard model' in physics regarding time dilation postulates that the lapsed time experienced by each observer depends on their movement relative to each other - the wrist watches of each observer will show different times (if measured accurately) because of their different movements relative to each other.

In fact every human being and every plane, train, car, boat and anything moving with a clock will show their own unique time on their cell phone or time-keeping device. If it were possible to measure time accurately, then according to relativity you would not find any two or more clocks with the same time. And furthermore, it is not just clocks that slow down with movement. Everything is said to slow down, your heart rate, thought processes and the movement of all the atoms in your body slow down.

Granted that time dilation at slow speeds is so tiny it can hardly be measured, if at all. Yet the special theory of relativity says that this is precisely what happens. Time dilation occurs at any speed, fast or slow, and it physically affects how time passes for any given object, living or dead.

If you take time dilation to its logical conclusion, it gets even crazier. According to Einsteinian relativity, movement of any sort, slow or fast, causes time to slow down ***and*** causes the moving object to physically contract in length (in comparison to anything not moving or moving more slowly).

Given that everything in the Universe is continually moving, it means that everything in the Universe is enjoying time dilation. Every animal, planet, star, grain of sand, and in fact every atom in the Universe is enjoying a slowing down of time and a contraction in its size by comparison to anything else. This of course is absurd, yet this is what Einsteinian relativity proposes.

In Einsteinian relativity it is clearly stated that time dilation is caused by movement at any speed, but that time dilation only becomes 'noticeable' at high speeds. By 'noticeable' it means capable of being measured by humans. The faster you go the greater time slows down. At the speed of light, time is said to stop.

The fact that light travels at a constant speed for all observers is no mystery. The constant speed of a train, a car or a caterpillar will also be the same constant speed for all observers. The constant speed of a car or caterpillar does not change just because it is observed from different vantage points, and the same goes for light.

Put another way, it is not possible to make the constant speed of light change its speed by moving towards or away from the light. Equally, it is not possible to make the constant speed of a car or a caterpillar change its speed by moving towards or away from the car or caterpillar. But of course human perception and imagination knows no boundaries, but that does not change what is happening in the real world.

The constant speed of something (including light) remains at the same constant speed irrespective of what observers may be doing, perceiving or imagining! Of course, if something gets in the way of the light, the car, or the caterpillar then their constant speeds can be affected.

Relative movement vis-a-vis time dilation

There has never been an experiment to show the veracity of time dilation for the simple reason that time dilation is completely baseless. Let us remember that time dilation means that time slows down with movement. So as you move faster, time slows down more and more. In other words, a moving clock slows.

It is asserted that time dilation is what proves special relativity. If the theory of time dilation falls apart, then like a house of cards both special and general relativity fall apart to be seen as spurious nonsense.

The theory of time dilation falls apart in the explanation that follows. You cannot be said to be moving unless you compare such movement to something that is not moving (that is at rest). Any kind of movement is not a movement unless compared to something at rest. So all movement is relative movement.

A car moves in relation to the ground that is at rest. The Earth moves around on its axis in relation to the Sun that is at rest. Electrons move around in relation to the nucleus of an atom at rest. So being at rest does not mean that the object at rest is not moving. It means that the object is at rest *relative to* a given moving object. Nothing in the Universe can be said to be moving except by comparison to something that is not moving, that is at rest.

Einstein asserts that you cannot compare your speed (movement) with an object at rest when the moving object is going through empty space. Put another way, the speed (movement) of a car cannot be compared to the empty space around it as a way of gauging its speed. Any kind of moving object requires another object so that the rate of movement can be compared between the two objects. This made

Einstein realise that ime dilation always had to apply to both the moving object and to the non-moving object against which the speed of the moving object is gauged.

For example, if rocket man A speeds past rocket man B, rocket man A can claim to be speeding at 300 k/h compared to rocket man B who is at rest. But it would be equally valid in the context of physics for rocket man B to claim that he is speeding at 300 k/h (albeit in the opposite direction) in relation to rocket man A who is at rest.

Einstein's special relativity predicts that both rocket men (A and B) will experience time dilation. Why both? Because you can argue (with equal validity) that both A and B were moving and that therefore time slowed down for both. 'A' was moving in relation to 'B' at rest. But equally, 'B' was moving in relation to 'A' at rest; therefore they were both experiencing movement and hence both experienced time dilation.

It has never been shown that this is so. Experiments using a moving clock in an aeroplane and a clock at rest on the ground have failed to show that the ground clock slowed down. Yet, the ground clock was moving in relation to the aeroplane clock (and vice-versa).

So the concept of time dilation becomes meaningless because all movement is relative. If a given object is enjoying time dilation by virtue of its movement, then how do we know it is moving? We know it is moving by comparing it against a non-moving object. But special relativity says that the non-moving object will also enjoy time dilation. Therefore there will be no difference in the amount of time dilation between the two objects, thus rendering the concept of time dilation as meaningless.

The take home message: time dilation is meaningless and non-existent.

The Majesty of Light

The underlying nature of light is widely misunderstood both academically and scientifically. In an effort to put this right, this section discusses the true majesty of light in the context of the *Final Theory of Everything*, and the widespread big misunderstanding of light is exposed for what it really is.

We will start by looking at the kind of statement that is often found when investigating the nature of light. A typical consensus quote is shown below (numbers 1 to 5 inserted for ease of reference).

Consensus Quote:

It's well known that light bends. (1) When light rays pass from air into water, for instance, they take a sharp turn; that's why a stick dipped in a pond appears to tilt toward the surface. (2) Out in space, light rays passing near very massive objects such as stars are seen to travel in curves. In each instance, light-bending has an external cause: (3) For water, it is a change in an optical property called the refractive index, (4) and for stars, it is the warping nature of gravity. (5) Researchers have shown that light can also travel in a curve without any external influence.

Unquote

This consensus quote is discussed as follows, under points 1 to 5:

1. Light going into water. It is often thought that light bends when going into water or some other medium. In fact, light can never bend under any circumstances, anywhere on Earth or in the Universe. This is so because of the very nature of light.

A light ray is a stream of photons. And a photon is a little self-contained packet of oscillating electricity and magnetism. A photon always moves at the speed of light (such speed denoted as 'c' in physics). Put allegorically, the photon is propelled forward by the 'leapfrogging' of electricity and magnetism. Each oscillation of the photon is a leapfrog. So when the electricity leapfrogs over the magnetism it lands a little beyond the magnetism. And equally, when the magnetism leapfrogs over the electricity it lands a little beyond the electricity. This continuous leapfrogging moves the photon forward indefinitely, at the speed of light.

Another fact is that light can only move in straight lines. For light to bend it would require the leapfrogging to bend. Put technically, it would require the transverse oscillations of a photon to bend and this has never been shown to happen. When light hits water it does not bend at all. This is what happens.

As soon as the photons in a light ray hit water, they hit atoms in the water. Such photons are absorbed (destroyed) by the electrons in atoms, and in their place new photons are emitted from electrons living inside the atoms. These new photons are called 'incident' photons because they are newly created photons. Incident light is simply a stream of incident photons. The science of how photons are absorbed, destroyed and then replaced by newly created photons is a well-studied subject, but very misunderstood.

Note: incident light is discussed in more detail in the next section 'The Virtual Video Camera'.

So when we see light 'bending' in water it is an optical illusion because the light that entered the water is destroyed. In reality we are seeing newly created incident light coming out of the atoms in the water, but at another angle. At no point has light ever changed direction; light can only ever move in straight lines. So when light enters water it does not take 'a sharp turn'.

2. Light rays out in space. The aforementioned consensus quote refers to light following a curved path as it passes massive objects in space such as stars. This implies that light is subject to the gravity of massive objects, but this is not so. In fact, light does not fall prey to any gravity, from anything. Light is completely unaffected by gravity. Here is some of the evidence.

2.1. Astronomers see that light travels in straight lines through the Universe. Here is a photo showing this:

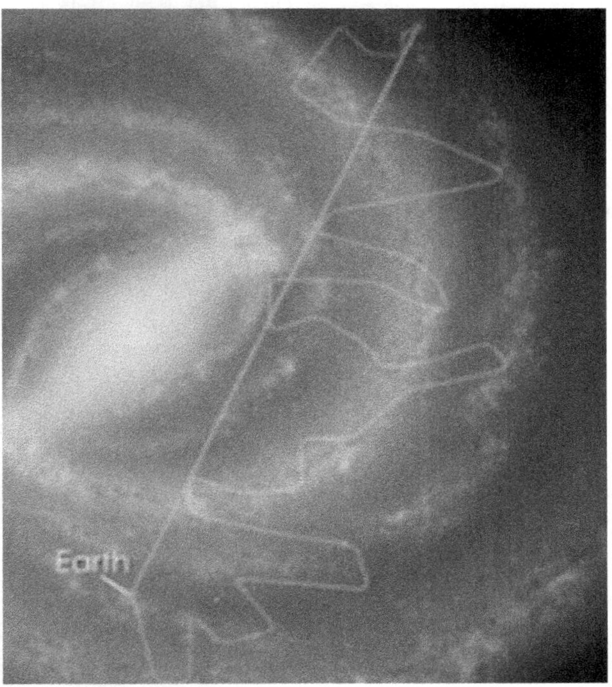

The above astronomical photo (credit to Nasa's Goddard Space Flight Center) shows a straight line for light and a meandering line crossing over the straight line for other types of cosmic rays going in the same general direction.

Commenting on the above image, Nasa said: *"Light travels to us straight from their sources* [straight lines], *but other types of cosmic rays are completely scrambled"* (source: Nasa Goddard Space Flight Center).

2.2 Another reason for thinking that light always travels in straight lines unaffected by gravity is the age of the Universe. It is about 13.8 billion years old. We know this by looking at light that has taken 13.8 billion years to reach Earth. But that light would never have reached us in its long journey through the Universe if it had been subject to the gravity of nearby objects it passed by. The light would have gone through or near many millions of stars and galaxies to reach us. It is inconceivable to think that for 13.8 billion years this ancient travelling light has enjoyed an empty highway straight towards Earth, completely bereft of nearby stars. This can only mean that light is not subject to gravity. This subject is also discussed in the section 'Cosmic Lensing'.

3. Refractive index. The third point in the above consensus quote refers to light changing direction or bending as it enters water: "*For water, it is a change in an optical property called the refractive index".* In fact, the so-called refractive index is very misunderstood. It is commonly believed that the refractive index *"is the measure of bending of a light ray when passing from one medium to another".*

This is not so because it implies that a light ray actually bends as it enters some medium (a very common misconception). The refractive index does indeed exist. It is simply an index (a list) of measurements. Each measurement refers to a different angle of light emission.

Thus, when light is absorbed and destroyed by virtue of hitting water or some other material, new incident light is emitted that replaces the destroyed light. This incident light (created by electrons in atoms) is emitted in straight lines, but at certain angles. The angle of emission is called the 'refractive index' - it does not indicate how much the light has bent, it indicates the angle at which newly created incident light is emitted. You can find a good description of refractive index in Wikipedia; it is a well understood science that we won't attempt to repeat here.

You may be wondering about fibre optic cables that carry light along all sorts of twists and turns. In fact, light never bends along a fibre optic cable. Here is an image (courtesy of howstuffworks.com) to briefly explain this:

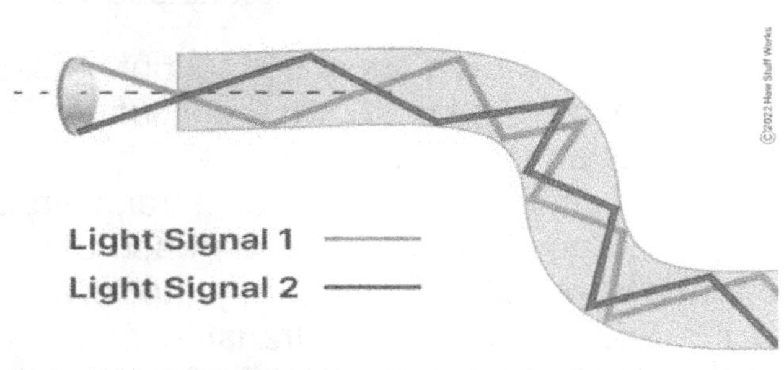

Take for example light signal 2 in the above image. When light is fed into the optic fibre cable it goes in a straight line and hits the inner side of the cable. Such light is then absorbed into the cable and new incident light is emitted outward (inside the cable) at the same angle. This new incident light also goes in a straight line and hits the inner side of the cable again further down, and so on as shown in the image. This is how light follows the bending turns of the cable without ever actually bending.

4. The warping nature of gravity. Point four in the consensus quote refers to the gravity of stars and its warping effect on any light coming its way. The implication here is that the gravity field of a massive object such as a star or a whole galaxy will affect or bend light as it goes by. But since light is not affected by gravity, it will not bend or be affected by any nearby object however massive. Light will simply continue on its way in a straight line.

Nevertheless, as light rays move forward they become less concentrated. A good analogy is to think of a growing cone of light. The face of the cone represents the face of the light moving forward shooting out light rays in all directions, in straight lines. So with time the density of light rays becomes thinner and thinner as the face of the cone expands. Any light reaching Earth from a great distance will be light that is much less concentrated than the light that left its source.

5. Light curving of its own accord. Last but not least, point five says: *"Researchers have shown that light can also travel in a curve without any external influence".* This is simply not so because light can only ever move as a transverse wave. It is well known in optic physics that with each oscillation the magnetism and electricity of a photon jump forward at right angles to the direction of energy transfer, thus galvanising light forward as a transverse wave, albeit always in straight lines.

The Big Misunderstanding of Light

Any talk of light actually curving or bending reflects a misunderstanding of the fundamental nature of light. Indeed, in science there is a big misunderstanding of light which is described in what follows.

As discussed, the electromagnetic oscillations of photons make light move forward at the speed of light. These EM oscillations are thought to make photons move as a vibrating sinusoidal wave, albeit always moving in straight lines. But it does not follow that the EM oscillations of photons determine the energy of light, hence the big misunderstanding. This misunderstanding has come to be enshrined in so-called 'lightwave theory' and is widely accepted in contemporary physics.

All photons in the universe are identical to each other. So every photon 'carries' the same amount of energy, namely the energy of one electromagnetic oscillation. Thus when the energy of light is referred to, a big misunderstanding arises. It is commonly thought that photons can have different amounts of energy, i.e. that the amount of energy can vary from photon to photon. This is false and has led to so-called 'lightwave theory', a theory that tries to justify the big misunderstanding.

The amount of energy in a ray of light is determined by the density (number) of photons in a given light ray. In other words, the energy of light is based on the number of photons in a given light ray, not the degree of energy carried by each photon.

To compound the big misunderstanding of light, lightwave theory stipulates that the rate (the speed) of electromagnetic oscillations determines the amount of energy in each photon. EM oscillations have nothing to do with the energy of light and everything to do with the movement of light. The rate of EM oscillations is the same for all photons; this sets the same universal constant speed of light everywhere in the Universe.

This big misunderstanding of light also leads to a misunderstanding of the frequency of light. The frequency refers to the concentration of photons in a light ray, i.e. how *frequent* they are. Such frequency is measured in a spectroscope that works out how many photons in a given light ray pass by a fixed point in one second of time.

The greater the frequency (higher concentration of photons), the smaller the distance between the moving photons. The lower the frequency (lower concentration of photons), the bigger the distance between moving photons. This ***distance*** between moving photons is known as the wavelength. Hence, the words frequency and wavelength are somewhat interchangeable.

The frequency (i.e. wavelength) tells you the colour spectrum. Different concentrations of photons show different colours. So different frequencies make spectroscopes (and make human eyes) see different colours. When we see a red car it means the light rays going from the car to human eyes have a frequency that matches the colour red.

But the mentioned big misunderstanding does not agree with this. Instead, it is believed that the frequency of light is determined by the rate (the speed) of electromagnetic oscillations of photons. For example, the quicker the oscillations the higher the frequency. But this is not so, there is no relationship at all between EM oscillations and the frequency of light.

Lightwave theory stipulates the following:

1. That the frequency of light does not change. It is set when light is created and does not change at any point. So when millions of lightwaves are created from different sources, each lightwave will have a different set-frequency that does not change. Even when light undergoes refraction (absorption/emission) it is said that the frequency of lightwaves remains unchanged.

2. That the rate (speed) of EM oscillations in a light ray is what determines colour, rather than the concentration (number) of photons in a light ray. Wave theory also defines light frequency as the measurement of the rate of EM oscillations in one second of time in a spectroscope. Note that in lightwave theory each EM oscillation equates to one wave of light and the number of such waves is said to determine the frequency.

This begs the question: *how does lightwave theory explain the colour spectrum of light?* Put another way: *how does lightwave theory explain how the length of a wavelength is determined?* If the frequency of lightwaves do not change as a result of going in and out of mediums and materials, then how do we see the millions of colour hues all around us?

This is a big contradiction that lightwave theory cannot explain, except to vaguely say that the length of a wavelength is determined by the Huygens-Fresnel Principle. But lightwave theorists are at a loss to explain exactly how this Huygen's Principle determines the length of wavelengths, and today the Huygen's Principle is largely discredited in optic physics. The eminent physicist Richard Feynman once said *"Actually Huygens' principle is not correct in optics"*.

Another contradiction of lightwave theory refers to the speed of light. If lightwaves can have different frequencies when created (i.e. different rates of EM oscillations), then how can the speed of light be the same everywhere?

And last, but not least, another contradiction of lightwave theory refers to the photoelectric effect. This well-known effect has been verified in many experiments. When light hits a solid material such as metal, some of the electrons in the solid material will be ejected completely from the material. Such ejection occurs immediately without any time delay because electrons tend to be thinly spread in solid materials and the nucleus of atoms have a weaker pull on their electrons. Hence, some of the electrons become 'overpowered' by the barrage of incoming photons and are ejected without first absorbing/emitting photons.

Wave theory cannot explain this photoelectric effect because it contradicts how wave theory is supposed to work. Wave theory says that when lightwaves hit a solid material such as a metal there has to be a tiny time-interval before any electrons are ejected from the material. Many experiments show that this is not so, and the contradiction remains.

Albert Einstein fully recognised the photoelectric contradiction faced by lightwave theory. Up to that point in his life Einstein believed in the wave theory of light, but upon recognising the contradiction exposed by the photoelectric effect he had to make a choice: either renounce wave theory altogether in favour of particle theory, or postulate that light has the characteristics of both. He chose the latter. From this moment onwards, the so-called wave-particle duality of light was born and to this day remains cemented in contemporary physics.

This fanciful and erroneous description of light-duality enabled lightwave theorists to explain away the photoelectric contradiction because it could be argued that in some circumstances light behaves as a wave, but in other circumstances (such as in the photoelectric effect) light behaves as particles, i.e. as a stream of photons.

Contemporary physics is increasingly showing the spurious nature of lightwave theory. Powerful computer-aided spectroscopy clearly shows that the frequency of light is not derived from the electromagnetic oscillations of photons. That the frequency of light is derived by calculating the concentration (the number) of photons in a given light ray.

The belief in lightwave theory is very widespread unfortunately, and it is holding back advances in optic physics. It is hoped that this book will help, however little, to redress the situation.

The take home message: Light is truly majestic and is one of the marvels of the Universe we inhabit. The true fundamental nature of light is based on streams of individual photons rather than streams of lightwaves. Each photon is a little packet of oscillating electromagnetic energy. The concentration of photons (not the rate of EM oscillations) determines the wavelength and frequency of light.

The Virtual Video Camera

This section reveals a Virtual Video Camera that is destined to revolutionise the exploration of the Universe more than anything else to date. It will galvanise our knowledge of cosmology and greatly increase the chances of discovering extraterrestrial life.

Attenuation of light

Light is an information carrier. The photons themselves that make up light do not and cannot carry information. But the time-intervals between the moving photons in a light ray do indeed carry information.

When light is created it disperses in all directions in straight lines. This dispersion takes the form of many streams of photons. The streams occur simply because trillions of photons are shooting out from the light source in all directions, so naturally many of those photons shoot out together as streams rather than as isolated single photons. But each photon in a stream is independent and self contained, and not coupled or joined to other photons.

A simple way to think of a light ray is to think of a stream of photons in a moving line. So every photon will have another photon ahead of it and another behind it, as the whole line of photons moves forward. In reality, in a light ray there will be many millions of such lines moving forward together, but not completely parallel, so that over time the lines of photons spread out making the photons in the light ray less concentrated.

Staying with the concept of a line of photons, there will always be a time-interval (a distance) between each moving photon. This time-interval will vary from light ray to light ray, and from one line of photons to another. This time-interval is caused by attenuation.

In physics the attenuation of light refers to the absorption of light into the atoms of objects it may hit, and to the emission of newly created light to replace the light received by such atoms. In short, attenuation is the absorption/emission of light. When 'virgin' light is created by the sun, a torch, a candle, a match, etc. this virgin light is easily and readily destroyed as soon as it hits an atmosphere, a wall, a medium or material of any sort. But such light is 'reborn' and emitted as incident light.

All the light we see around us is incident light, i.e. light that has been attenuated. When light is absorbed into an object, the electrons in the atoms of the object absorb and destroy or change the received light. It is gone forever as light. In doing so the electrons absorb the energy of the destroyed photons. The electrons are then compelled to get rid of the absorbed energy, and they do it by releasing brand new photons of the same energy.

All photons are made this way - they are created by electrons. In the case of 'virgin' light, the electrons in a light bulb, a candle or the sun are excited (over-charged with energy) in various ways, compelling such electrons to release photons.

The mentioned time-intervals between moving photons are caused by attenuation. The time it takes for an electron to absorb and then emit a new photon varies tremendously depending on the type of object, material or medium. This time-interval can therefore be a short or a long time-interval. The length of the time-interval will determine the degree of concentration of photons in a given light ray. The shorter the time-interval, the greater the number of photons in a given light ray.

The Information Carrier

Now we come to exactly how light carries information. No photon itself carries information, but the time-interval between moving photons does indeed carry information. For example, when a red car sends light rays to your eyes - the time-intervals in such rays will match the colour red and tell your brain that you are seeing the colour red.

Different time-intervals (caused by light attenuation) match different hues of colour in the overall colour-spectrum of light. Thus, in this example, the light rays going from the red car to your eyes have carried information pertaining to the colour red. Any and all information carried naturally by light comes solely from the time-intervals between moving photons. This is how light carries information.

Spectroscopes work in a similar way. When light goes into a spectroscope, such light is analysed to determine the time-intervals between moving photons. More specifically, the spectroscope works out the wavelength in a given light ray, and hence the frequency of such light. All this comes from the time-interval (or distance) between moving photons. The greater the time-interval, the greater the distance between moving photons, and hence the longer the wavelength.

The length of the mentioned wavelength tells the spectroscope the colour pertaining to the wavelength. And the colour hue can also convey a lot of other information that can reveal types of chemicals and materials, and even types of atoms. Scientists have created very detailed 'colour charts', and each colour refers to certain materials, chemicals, etc. So just by knowing the colour hue of a light ray, scientists can deduce a wide gamut of information.

Thus light can only convey a particular colour (no other type of information), but that is enough to convey a host of other information ranging, for example, from the type of soil on the ground to the type of atmosphere above the soil.

For example, when a planet receives starlight from its nearest sun, such light will be 'reflected' (attenuated) out into space as incident light. If we can see such light on Earth we can glean much information from the light such as the composition of the soil, the planet's atmosphere, presence of water, and so on.

The Camera

Given the fundamental nature of light as revealed in this book and given the current level of spectroscopy science, the following prediction is offered as a matter of speculation:

It is predicted that by about the year 2035 it will be possible to see close-up moving video images in full colour and sound, coming from the surface of any planet or star, however faraway the objects may be. Here is why this is theoretically and technically possible:

1. A story to tell. Providing that light from a given planet can be seen on Earth, it means that such light will be incident light garnered from its nearest star. Thus many millions of streams of incident light will be arriving continuously, and each incident ray of photons will have a different frequency and a different story to tell.

2. A wealth of information. Each of the many millions of incident rays arriving on Earth will have a different frequency depending on where the incident light is emitted from. Many different incident rays coming from the same planet or star, will provide a wealth of information.

3. Artificial intelligence. Future spectroscopy will become much more advanced on the back of more powerful computers and artificial intelligence. Thus, such spectroscopes will be combined with very powerful computing and AI capabilities that will be capable of analysing and compiling the many different frequencies of incident light received from a given planet or star.

4. Virtual video camera. Given future developments in spectroscopy, it is theoretically and technically probable that the millions of different light frequencies received will be converted into a panorama of full colour, just as we humans are doing every day when we look at things on Earth. This means that through the virtual video camera we will be able to see things on the distant planet in full colour and shape, just as if we were standing on the given planet looking around at things. It will be a virtual video camera that will be possible to 'put' on the surface of any planet or star that sends us incident light.

5. Just like a movie. The many millions of images compiled by future spectroscopy (as described in this prediction) can then be compiled or converted into moving images. This is a well-understood science given current digital cinematography and video technology. It will be a small step to apply such knowledge to the many millions of images extracted from incident light received from a given planet or star. Let us remember that human eyes can already process millions of different light frequencies to end up seeing the full colour and movement of things around us. We are nearly at the point of being able to do the same with powerful spectroscopy as it develops in the future.

6. Stars and planets. Thousands of exoplanets have been identified by astronomers and they all emit incident light arriving at Earth. But much of this incident light is drowned out by the planets' local stars. However, the detection rate of such incident light is greatly improving as technology improves.

There will be no shortage of distant planets that we can explore, using advanced spectroscopy that will give us full high-definition video images of such planets. The same goes for starlight - incident light from stars can be analysed the same way as incident light from planets, showing the composition, age and other factors of a star.

For high quality definition, free of distortion from the Earth's atmosphere, relay stations can and will be set up in the form of one or more satellites. Incident light from a given planet or star could be received by a satellite relay station and beamed to Earth free of atmospheric distortion.

7. Distance no object. Putting all this together, it is predicted that by 2035 we will be able to see full video images as if we had a video camera stationed on a given planet. If the incident light from a planet that is, say 100 light years away, it means we will be looking at video scenes that occurred on that planet a 100 years ago. Distance is no object because the incident light from that far-away planet is already continuously arriving on Earth.

8. Adding sound. It is also predicted that it will be possible to add sound to the mentioned video images. It is not suggested that incident light will somehow be able to record and carry audio waves to Earth. But using the ingenuity of AI and very powerful computer technology it will become possible to incorporate sound into the silent video images, giving a very realistic sound-track to such videos. This will give us a full virtual video, coming to Earth from distant planets.

Adding sound to silent movies for example can already be done, and the groundwork for adding sound to scientific endeavours is well underway. *"Astronomy is often thought of as a visual science that produces stunning images of the cosmos, but it's possible to hear it as well"* (source: Patchen

Barss, How sound is providing new clues about the Universe, Oct 2023, bbc.com).

Advanced technology will be able to take clues from the silent video of a far-away planet, such as dust movements, changing scenery, weather, and any kind of movement, so as to add a realistic sound-track to the video.

9. A revolution in cosmic discovery. Future spectroscopy technology as predicted here will revolutionise astronomy in unforeseen ways and tell us things about other planets and stars that we cannot do today, even with super-powerful telescopes. The search for extraterrestrial life and the way other planets have evolved are just two of the many things that come to mind.

10. A full colour audio-visual movie. High quality optic resolution of the virtual video images will be achieved with artificial intelligence and powerful video enhancing software which will be capable of greatly improving the resolution of such images. Thus, the end result will be high quality, full colour & sound videos (just like a movie) of distant planets and stars as if we had put a physical movie camera on the surface of a planet or star.

In a sense the predicted virtual video cameras already exist and are already in place on planets and stars, and we are already receiving their video images on Earth. But we humans do not yet have the capability to detect these video images. For example if a planet is 50 light years away, we will be able to 'put' a virtual video camera on that faraway planet with no time delay at all because the light rays arriving on Earth already 'contain' such video recordings, albeit they will show us things 50 years old in this example.

Scientists are urged to take up the challenge of developing advanced spectroscopy as predicted here in this book. This has the potential to revolutionise our understanding and exploration of the cosmos, probably more than anything else.

The take home message: Optic physics and computerised spectroscopy is advancing rapidly. This, combined with artificial intelligence, will give us virtual video cameras that can be instantly ‘put’ on the surface of any planet or star that sends us their light. This will revolutionise cosmology and the exploration of the Universe more than anything else to date.

Cosmic Lensing

Cosmic lensing is real, but so-called 'gravitational lensing' is not real. To be clear, the lensing effects that are commonly used and seen in the Universe are real and they enable astronomers to see distant objects as if looking through a giant magnifying glass.

Here is an image of cosmic lensing:

In the above image we see from left to right that light from a distant galaxy travels in straight lines towards Earth by virtue of being in the same line of sight. But such light hits a cluster of galaxies that's in its path. When this happens the cluster of galaxies in the middle absorbs some of the light and emits new incident light due to refraction.

But the very large size of the cluster of galaxies acts like the concave shape of a magnifying glass. When we look at this cluster of galaxies, we of course see the objects closest to Earth at the forefront of the cluster, thus giving the cluster an outward concave shape. This concave shape acts like a giant magnifying glass.

If you're wondering how it can be possible for Earth to see a distant galaxy that is directly behind a cluster of galaxies, here is an explanation. In the above image we see incident light going from the distant galaxy (left to right in the image) to the cluster of galaxies in the middle. When this incident light arrives at the cluster of galaxies, this cluster will be full of 'holes' (spaces between objects and galaxies). So the mentioned incident light continues on its way through the holes towards Earth.

So astronomers on Earth receive incident light directly from the distant galaxy, and they also receive incident light directly from the much nearer cluster of galaxies. Put simply, two lots of streams of light are received, one from the distant galaxy and one from the cluster. Both these two lots of streams of incident light are magnified and hence we see a magnification of both the distant galaxy and the nearer cluster of galaxies.

Note: It is known that stars and planets have 'atmospheres' composed of solar winds and various types of gases and dust. This means a whole galaxy will have a strong refraxing effect on any light coming its way. And equally, any light emitted by a galaxy will be virtually all incident light.

To clarify this point further, the photons that make an incident ray of light move at the same constant speed of light. But the journey-time of a ray of light *as a whole* will depend on the time-interval between the moving photons. So each photon in a ray of incident light never slows down. But a light ray as a whole, with long time-intervals between photons, will take longer to arrive at, say, Earth compared to a light ray with shorter time-intervals.

When this incident light reaches Earth it gives astronomers information of the objects it has come from. This information is derived from the frequency of the light received. But in the case of cosmic lensing, the magnified picture we see is always a little fuzzy and distorted due to chromatic aberrations.

When many streams of refracted light travel at different speeds in nearly parallel lines to reach a spectroscope (or a camera) it can cause chromatic aberrations in the magnified image being received. Also, chromatic aberrations can be caused by the very uneven 'surface' of the cosmic magnifying glass, making some objects appear in two or more *apparent* locations instead of just appearing in one *real* location.

"Chromatic aberrations are caused by different wavelengths moving at different speeds through [for example] *media that is denser than air"* (source: Chromatic Aberrations, telescope optics.net).

By receiving cosmic lensing images at a receiving station that is outside Earth's atmosphere, the lensing images will be unaffected by the refraction effects of Earth's atmosphere. The images can then be relayed to Earth with much greater clarity.

Cosmic microlensing (known mistakenly as 'gravitational microlensing') is part of the cosmic lensing phenomena but on a smaller scale. Like cosmic lensing, microlensing works through light refraction and diffraction, not through any kind of gravitational effects.

"Microlensing magnifies the distant source, revealing it or enhancing its size and/or brightness. It enables the study of faint or dark objects such as brown dwarfs, red dwarfs, planets, white dwarfs, neutron stars, black holes, and massive compact halo objects. Such lensing works at all wavelengths, magnifying and producing a wide range of possible warping for distant source objects that emit any kind of electromagnetic radiation" (source: Gravitational Microlensing, Wikipedia.org).

Clearly Wikipedia is incorrect to attribute cosmic and micro lensing to the effects of gravity, and it is hoped that with time this will be corrected.

All the various types of lensing phenomena that astronomers experience (including eclipses, Einstein rings and microlensing) can be fully explained by cosmic lensing and/or light refraction rather than by gravitational light-bending. The solar eclipses of 1919 and 1922 that were deemed to verify Einsteinian relativity can be simply explained by light refraction and diffraction.

Gravitational lensing postulates that *"a gravitational lens can occur when a huge amount of matter, like a cluster of galaxies, creates a gravitational field that distorts and magnifies the light from distant galaxies that are behind it but in the same line of sight. The effect is like looking through a giant magnifying glass"* (source: Gravitational Lensing, Nasa, 2022).

Gravitational lensing (i.e. lensing caused by gravity) is spurious for several reasons:

If gravitational lensing accounts for the magnified images that astronomers see, then such images would not be subject to chromatic aberration or fuzzy images - they would be distortion-free because although the light bends, it is not refracted or diffracted. But in reality such light is not distortion-free. The following image shows how gravitational lensing is claimed to work (image courtesy of Getty Images):

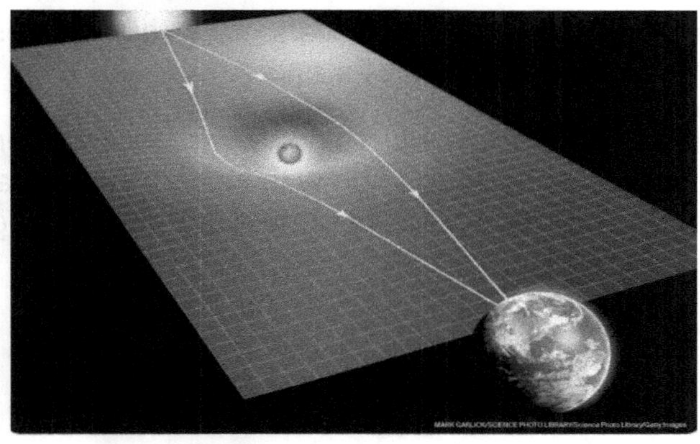

The above image shows how a curvature of space would make light split up and go round, say, a cluster of galaxies (in the above image the cluster is shown as an object pushing down on a trampoline between the two rays of light). Having gone round the cluster of galaxies (or some massive object), the split light reaches Earth from two directions. It is claimed that these two directions of received light are what causes a magnifying effect.

But how exactly is such magnification achieved? A typical explanation goes like this: *"Gravitational lensing occurs when a massive celestial body — such as a galaxy cluster — causes a sufficient curvature of spacetime for the path of light around it to be visibly bent, as if by a lens. The body causing the light to curve is accordingly called a gravitational lens".*

This does not explain at all how the actual magnification phenomenon occurs. At best, spacetime curvature may explain how the same object may appear in two apparent locations of space as a result of the split light, but it does not explain any kind of magnifying phenomenon.

In reality, the movement of light is only ever rectilinear. It does not split up and go around objects from both sides, and light does not follow any kind of bent path through spacetime curvature. Light only ever travels in straight lines, never bending and never changing its angle of travel.

The phenomenon of magnification is very well understood in science, and it has nothing to do with spacetime curvature. It's entirely to do with refraction as shown in the following image:

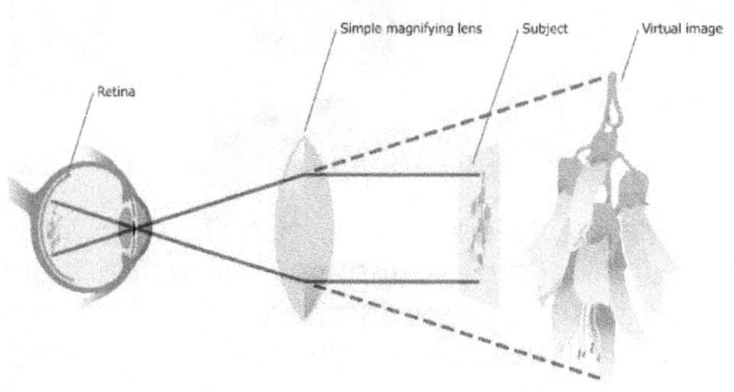

This image shows that light from an object (from right to left) passes through a biconvex lens and is bent (refracted) towards your eye. It makes it appear to have come from a much bigger object. The light is not split at any point - it goes into the magnifying lens and is then absorbed and emitted (i.e. refracted) at a focused angle towards the human eye or telescope. The dotted lines in the above image show the perceived magnified image. It's the same with cosmic lensing.

Gravitational lensing is entirely dependent on light having enough mass to cause spacetime curvature, but it has never been demonstrated experimentally that light has enough mass to be affected by gravity. Experiments across the surface of Earth have always resulted in light travelling in a straight path and not falling prey to any kind of gravity.

However, it is thought that light may have a very tiny amount of mass arising from its kinetic energy, but not enough mass to be affected by the gravity of cosmic objects that it may pass by.

Is there any evidence that light is not affected by the gravity of nearby objects as it travels through space? Yes, evidence comes from much astronomy research showing that light always travels in straight lines, whereas other types of cosmic rays containing a variety of subatomic particles do not. In a previous section of the book (the Majesty Of Light) we saw a photograph of light travelling in a straight line through the cosmos. And the fact that ancient light from distant galaxies is able to reach Earth shows that such light was not waylaid by the gravity of millions of stars and galaxies it went by.

Further evidence that light travels through space unaffected by gravity is shown in the following image:

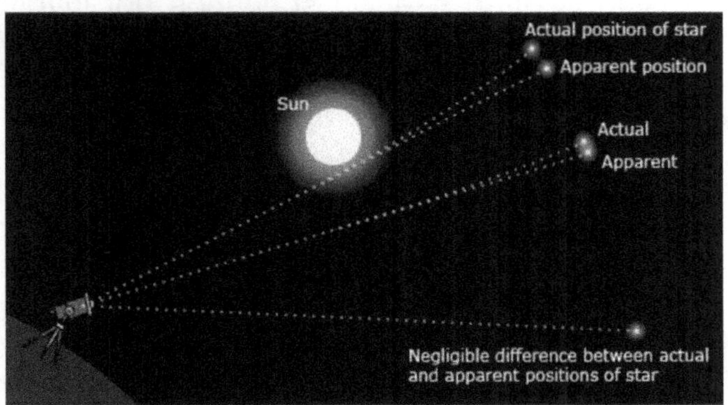

(Image courtesy of Gatot Soedarto, Deflection of Light by Refraction, Not Gravity)

In the above image as the starlight ray goes from right to left close to the sun on its way to Earth, spacetime gravity would bring the passing starlight closer to the Sun. For an observer on Earth, this would increase the distance between the actual and apparent position of the star emitting such starlight. The further away the passing starlight ray from the sun, the smaller the distance between actual and apparent positions.

The fact that this does not happen shows that the Sun's gravity is having no effect on the passing starlight. The actual and apparent positions of stars in our night sky are not affected by the Sun's gravity whether it be during the day or during the night. The above image therefore rightly depicts the falsehood of spacetime gravity.

Relativists may argue that the Sun's gravity is not strong enough to affect light. That it takes the huge gravity of a whole galaxy or cluster of galaxies to have a gravitational effect on light. This is a false argument because spacetime gravity is not a force. Einstein himself was adamant that spacetime gravity is not a pulling or pushing force at all. And furthermore, cosmic lensing is often seen in scenarios not involving the intervention of a whole galaxy or cluster of galaxies.

If spacetime gravity existed then a huge mass such as a galaxy would no doubt have a big effect on the curvature of space, akin to making a big dent on a trampoline. But if light does not have enough mass to make any significant dent on that trampoline, then in this fanciful example the light will not move closer to the galaxy it is passing, however strong the gravity of the galaxy. Put another way, the strong gravity of the galaxy and the insignificant gravity of the light are such that they are not able to detect each other on the trampoline of spacetime, so light continues on its way unaffected.

The point here is that light travels through the Universe unaffected by the gravity of nearby objects during its journey. It is therefore very unlikely that light would be affected by so-called gravitational lensing. Just the fact that scientists on Earth can receive light that has travelled for billions of years shows that such light has travelled in straight lines without falling prey to the gravity of the many objects it must have passed by. And if light is not affected by gravity, it means that the theory of gravitational lensing and indeed the general theory of relativity is spurious.

All empirical and experimental evidence shows that light cannot follow a bent path as envisaged in spacetime curvature. It may be argued that light continues in a straight line as it travels through a curvature of space, but this is disingenuous to say the least. How can any object follow a completely straight path and at the same time follow a bent path so as to go towards another object?

Cosmic lensing then is what is used by astronomers to map out and gain a better understanding of the Universe. By understanding that gravitational lensing is spurious and that cosmic lensing is the way it works, it better equips scientists to refine and improve their cosmic lensing strategies into the future.

The take home message: Cosmic lensing is a marvel of nature that helps scientists understand and map the Universe. Gravitational lensing is spurious because light is not subject to gravity, and because the magnification effect is caused by refraction rather than by gravity.

The doppler effect

In this section the doppler effect of light is discussed and the redshift is explained showing that distant galaxies are receding from Earth in all directions.

The doppler effect is very much in our daily lives in various ways. For example, the change of pitch heard when a vehicle sounding a horn approaches and recedes from an observer. Or the doppler shift that causes a rainbow when sunlight shines through rain drops. In the context of cosmology, when a source of light such as a distant galaxy is receding from Earth (due to cosmic expansion) we will see a redshift occur in the spectrum of such light.

It is well-established that a longer frequency of light (i.e. flatter, less bunched up photons) creates a shift to the red end of the light spectrum when gauged through a spectroscope. But what exactly causes light to become a longer flatter frequency when coming from distant cosmological objects?

Relativists would argue that so-called lightwaves become flatter due to time dilation. The following image shows the crests and troughs of an imagined lightwave.

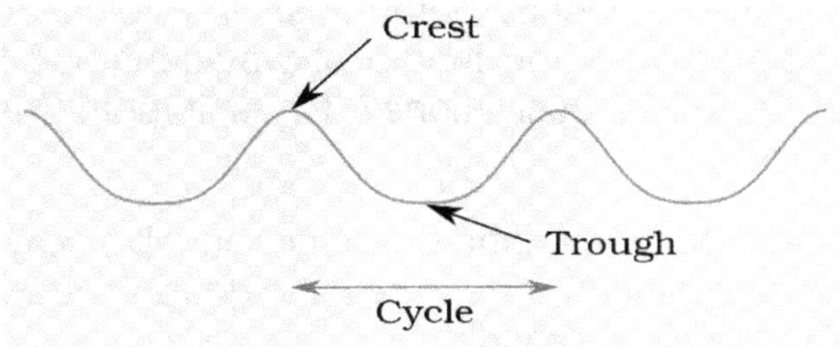

Given that the speed of light is constant, if it takes more time for light to travel from each crest to each trough, the wavelength will become longer (flatter, less bunched up) as in the following image:

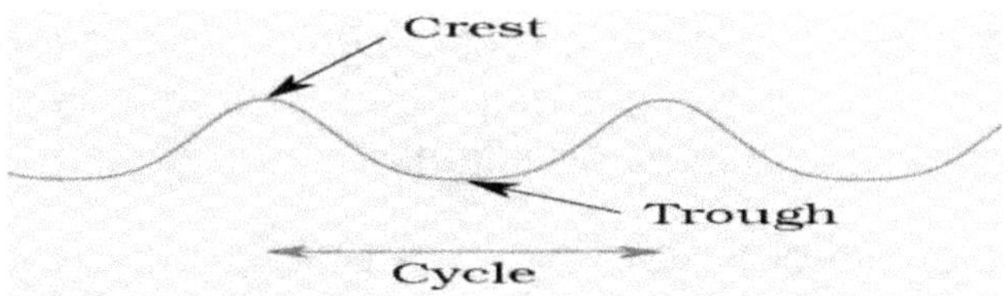

So the speed of light does not change, but relativists argue that the longer wavelengths are caused by time dilation. That is, due to time dilation it takes light more time to travel from each crest to each trough (because time dilation slows down time). And the longer the travelling time, the longer and flatter the lightwaves. And this, it is claimed, is what causes a redshift.

This raises a big contradiction. If the length of wavelengths is determined by time dilation, then how are colours perceived? In wave theory it is stipulated that the length of wavelengths is set by the rate of electromagnetic oscillations. And that the rate of EM oscillations in photons are set at the moment that light is created (or at the moment light is emitted).

So if the length of wavelengths determines the frequency of light, and if the frequency determines the colour spectrum, then what determines the length of wavelengths? Is it time dilation or is it EM oscillations? Or is it both (hence the contradiction)?

The doppler effect of light does not involve any kind of relativity or time dilation. This is how the doppler effect actually works with regards to redshift:

1. Light source recedes. The AE force (Accelerating Expansion force of the Universe) moves distant stars and galaxies away from Earth as the Universe expands in all directions. So from our perspective, when we look through a telescope and see a distant galaxy receding, its light is also receding along with the receding object. The further away an object, the faster it recedes from Earth.

2. The red shift. As the object recedes, it means its light is taking longer and longer to reach us with each passing moment. So although the light comes to us at the speed of light, it is nevertheless receding all the time.

Put simply, imagine that a photon takes off from the receding object on its journey to Earth. The next photon behind it also takes off towards Earth, but when the second photon takes off, the object has already moved further away from Earth. And so on with the next photon, and the next. As a result the stream of photons coming to Earth will have time-intervals between the moving photons that are becoming longer and longer.

This will show up in a spectroscope as a frequency that is becoming longer and longer as each photon passes through the spectroscope. Unlike incident light which has a relatively uniform frequency, the light from a distant receding object will have a frequency that is gradually and continually changing. This gradually changing frequency, that is becoming longer and longer, is what shows up as a redshift. The light is *shifting* more and more towards the red end of the light spectrum so we call it a redshift spectrum rather than a red spectrum.

The transverse doppler effect

The transverse doppler effect, according to Wikipedia.org, refers to *"the blueshift predicted by special relativity that occurs when the emitter and receiver are at their points of closest approach; or the redshift predicted by special relativity when the receiver sees the emitter as being at its closest approach. The transverse Doppler effect is one of the main predictions of the special theory of relativity".*

Put simply, the above means that if a source of light is moving at right angles to (transverse to) the line joining an observer to a light source, the observer will see a change in frequency or wavelength, even though the distance between observer and source at that instant is not changing. It is claimed that this proves time dilation.

The notion of the transverse doppler effect arose from a research paper titled 'Michelson-Morley Null Result'. In his research Michelson made a false assumption concerning the angular path of the transverse light pencil. More specifically, he made a false assumption about the lateral inertial and angular propagation of a perpendicular light ray, thus leading to a false hypothesis called the transverse doppler effect. This false hypothesis was adopted by Einstein and his followers, incorporated into Einstein's relativistic moving light clock theory, and asserted as an experimental confirmation for the validity of Special Relativity.

Experiments such as Ives-Stilwell and others have tried to prove the validity of the relativistic and transverse doppler effects. They were born from the baseless transverse doppler effect and the Michelson-Morley experiments. The scientific community of today has shown conclusively that the Ives-Stilwell experiment and other similar tests such as Michelson-Morley and Kennedy Thorndike are flawed, mainly because they assumed the existence of an all-pervading ether.

Here are some of the many commentaries on this matter:

"Based on recent theoretical findings of the relativistic transverse Doppler effect involving the millennium theory of relativity it can now be shown that such comparison between longitudinal and transverse effects is fundamentally flawed and subsequently invalid because it assumes compatibility between two different mathematical treatments where under

the conditions of the experiment, none exist. More specifically, it can now be shown that the special relativity mathematical treatment of the transverse Doppler effect is invalid and thus incompatible with the longitudinal mathematical treatment at distances close to the moving source. Subsequently, any direct comparisons between the longitudinal and transverse mathematical predictions under the specified conditions of the experiment are invalid". Source: Ives – Stilwell Experiment Fundamentally Flawed, http://www.mrelativity.net/, 2007.

*

"Are we wrong about the Michelson Morley Experiment? The answer is yes. The light-clock, a popular way to explain to students the concept of time dilation, doesn't work. The Theories of Relativity are based on a wrong skewed line in the drawing of the Michelson-Morley experiment" (source: Hans Deyssenroth, Are We Wrong about the Michelson Morley Experiment? J Phys Math 2020, 11:1).

*

"The Michelson-Morley experiment has been described as the greatest negative experiment in the history of science because, although Michelson refused to admit it, the experiment failed to prove that the ether existed" (source: Loyd S. Swenson Jr., The Ethereal Aether: A History of the Michelson-Morley-Miller Aether Drift Experiments, 1880-1930, University of Texas Press, 1972).

The take home message: In cosmology the doppler effect can be seen as a redshift. This redshift is caused by a journey-time of photons that becomes longer and longer to arrive on Earth because such photons are departing from a receding object. The doppler effect is not caused by relativity.

Faster than light

This section explains why light can indeed go faster than the standard speed of light at c. And why Einstein's concept about the speed of light was simply wrong.

There is no mystery about the speed of light being the same for all observers. When it is said that the speed of light is constant (if nothing gets in its way) for all observers, it is an oxymoron. If the speed of light is unchanging for observer A, then of course it will also remain unchanging for observer B whatever his location or movement. And it's the same for anything else moving at a constant speed!

For example, if a train is moving at a constant speed of 100 km/hour, the speed continues to be the same in relation to the ground whether observed from a train station platform or from a car driving alongside the train at the same 100 km/hour. When you look out of the side window of your car at a part of the train, that part of the train (e.g. that train window, say) may appear to be stationary in relation to the car, but of course the speed of the train has not changed in reality. The speed of the train is related to the ground on Earth and does not change however the train may be viewed from inside or outside.

If you were to drive towards the train and pass it on a parallel road, the train will be perceived to be going faster than 100 km/hour. But this is just human perception, the speed of the train has not changed (it remains 'constant').

The same goes for light. Imagine that you're in a spaceship and you see a beam of light travelling through space. Now imagine that you speed up your spaceship and travel alongside the beam of light at the same speed. When you look out of the side window of your spaceship at a particular bunch of photons they will appear to be stationary in relation to your spaceship, but of course the speed of the light has not changed in reality.

If you were to travel towards the beam of light in the spaceship and pass it on a parallel trajectory, the beam of light will be perceived to be going faster than the constant speed of light. But this is just human perception, the speed of the light has not changed (it remains ‘constant’).

Can the speed of light be exceeded? Absolutely. Let’s say we have two scenarios. **Scenario A:** you drive a car at 100 kph and a passenger in the back seat shines a torch forward as you drive. **Scenario B:** Same as scenario A, and as you drive you pass a person standing beside the road also shining a torch in the same direction as your car’s movement.

In scenario A the speed of the torchlight is speed c plus the car’s speed of 100 kph. In scenario B the speed of the torchlight is just speed c. So in scenario A the photons are moving at speed c plus 100 kph relative to the road. In scenario B the photons are moving at just speed c relative to the man standing in the road.

Using the above example, Einstein was wrong to say that the speed of light relative to the road is the same in both scenarios A and B. He was wrong to say that the photons being carried in the car at 100 kph go at the same speed relative to the road as the photons emanating from a torch held by a man standing in the road.

But to qualify this, the speed of light cannot be exceeded relative to the inertial frame in which the force is being applied. In other words, the torchlight in scenario A will only ever travel at speed c relative to the passengers inside the car.

Here’s a practical example that anybody can do. Suppose you’re sitting in a train going at 100 kph. You then get up and walk forward at 5 kph inside the train. This means you are moving at 105 kph in relation to the ground outside. So you have gone faster than the train without affecting the train's speed at 100kph. The train is not endowed with some mysterious power that prevents another object going faster. Equally, light is not endowed with some mysterious power that prevents another object going faster.

Einstein's concept of the speed of light is wrong because it is based on comparing moving light with a non-moving object such as the road, the Sun or Earth. This led Einstein to reason that the difference in speed between the moving light and a non-moving physical object is what determines the speed of light as being constant at speed c. But this is incorrect and shows that Einstein did not understand or did not read Maxwell's 1865 treatise.

Let's pose the question: *what exactly makes the speed of light constant and the same everywhere?* In contemporary physics, the relativity-answer goes like this:

"The speed of light in a vacuum is determined by the properties of space and time, as described by the theory of special relativity. The speed of light is a fundamental constant of nature and is the same for all observers, regardless of their relative motion. Special relativity shows that the speed of light is a universal constant that emerges from the symmetries of spacetime and the nature of electromagnetic interactions".

So put simply, relativity is saying that the speed of light is constant and that such constancy arises from spacetime and the electromagnetism of light (this is meaningless and spurious). Also, relativity is saying that the speed of light is the same for all observers regardless of their speed relative to the light being observed (also meaningless and spurious).

Relativity fails to explain exactly what makes the speed of light constant. On the other hand, James Maxwell (1831-1879) correctly describes the speed of light as being constant in a vacuum, and correctly gives the formula for calculating the speed of light (about 300 million metres per second). The phrase 'in a vacuum' simply means nothing getting in the way of the light as it moves.

Maxwell understood the nature of electromagnetism and deduced correctly that the oscillating electromagnetism of photons acted to propel the photons forward at the speed of light. But in the 1800's it was not known how electrons create light. It was not known that electrons, because of their nature, always create light at the same speed c. Even when a photon is absorbed into an electron, the electron is obliged to create a new replacement photon of exactly the same energy, and hence the same speed of light.

So although Maxwell could not explain exactly what determines the constancy of the speed of light, he was on the right path when he published his theory of light in 1865. About 40 years later Einstein published his particular version of the theory of light referred to as the theory of special relativity. But it seems that Einstein did not understand (or did not want to copy verbatim) Maxwell's theory of light. Unfortunately, like many others, Einstein fell victim to the big misunderstanding of light (see the section *The big misunderstanding of light).*

So what makes the constant speed of light constant? The answer lies in the nature of atoms and their electrons, and the principle of the conservation of energy. This principle is well-known to science; it basically says that energy is neither created or destroyed (only changed). In the case of atoms, they can take on more energy temporarily, and then such energy must be released such that the atom is left with the same amount of energy as before. Failing this, the atom is destroyed or changed into something else.

Thus when electrons inside atoms are excited (given extra energy) they are compelled to release the extra energy in the form of photons and then return to their previous state of energy. This extra energy in atoms is how light is created from sources such as the Sun or by striking a match. Regarding incident light, it is created when the electrons in atoms absorb photonic energy and then release the absorbed energy in the form of newly created photons.

When electrons do this they release exactly the same amount of energy as the extra energy received, leaving the electron with the same amount of energy as before (the conservation of energy principle at work). By always creating or emitting photons with the same amount of energy, it ensures that such photons always travel at the same speed of light everywhere.

When an electron creates a photon it is creating a little field of electromagnetic energy that immediately starts oscillating and moving off at the speed of light. The energy of a photon is the energy of one electromagnetic oscillation which is enough to 'kick start' the EM oscillations into action sending the photon on its way. This energy has nothing to do with the various types of electromagnetic energy that we see for example in X-rays, gamma rays, sunlight rays, radio waves etc. These various types of EM energy are entirely determined by the concentration (the number) of photons in a given light ray.

As to why atoms and electrons must always observe the principle of energy conservation, the answer is that it is a fundamental principle of nature. It is the way the Universe works and has nothing to do with spacetime or relativity.

The take home message: Once we free ourselves of Einsteinian relativity it becomes easy to understand how it is possible to go faster than the speed of light and exactly what determines the constancy of the speed of light.

The Fundamental Forces of the Universe

This section reveals the nature of the fundamental forces of the Universe and how they work in concert to give us the Universe we inhabit. This sets the scene for revealing a fifth fundamental force of nature.

In the current standard model of physics there are four fundamental forces at work in the Universe: the Strong force, the Weak force, the electromagnetic force, and the gravitational force. They work over different ranges and have different strengths.

The Strong and Weak forces

The Strong and Weak nuclear forces are caused by movement. Put simply, certain movements in subatomic particles cause changes to those subatomic particles, and from those changes the Strong & Weak forces are borne. Technically, the strong force changes quarks into protons, neutrons, and other hadron particles, and binds them into the nucleus of an atom. The weak force changes protons into neutrons and vice versa.

So the Strong and Weak nuclear forces cause changes in subatomic particles, and this in turn affects many aspects of the micro and macro Universe. These changes occur because of movement and interaction between subatomic particles.

As the Universe expands at an accelerating rate it puts more space between things. But if things are close enough (such as the atoms in our body, or the planets in our solar system) the 'closeness' trumps the pressure to separate exerted by the expansion effect. Thus, even though the expansion effect is not enough to physically separate subatomic particles, it is nevertheless enough to exert a resistance to separate. It is this resistance to separation that excites subatomic particles, making them move more quickly.

This increase in movement is what causes the Strong & Weak forces to come about and affect subatomic particles, each in their own different way.

Gravity is the weakest of the fundamental forces, but it has a global (universal) range. The electromagnetic force also has a global range but it is many times stronger than gravity. The Strong & Weak forces are effective only over a very short range and dominate only at the level of subatomic particles.

These four fundamental forces act upon us every day, whether we realise it or not. From playing basketball, to launching a rocket into space, to sticking a magnet on your refrigerator - all the forces that all of us experience every day can be whittled down to a critical quartet: Gravity, the weak force, electromagnetism, and the strong force. These forces govern everything that happens in the Universe.

Here is an abridged extract from Wikipedia.org, 'Fundamental Interaction'.

Quote:

In physics, there are four fundamental interactions: the gravitational and electromagnetic interactions, which produce significant long-range forces whose effects can be seen directly in everyday life, and the strong and weak interactions, which produce forces at minuscule, subatomic distances and govern nuclear interactions. Some scientists hypothesise that a fifth force might exist, but these hypotheses remain speculative.

Each of the known fundamental interactions can be described mathematically as a field. The gravitational force is attributed to the curvature of spacetime, described by Einstein's general theory of relativity. The other three are discrete quantum fields, and their interactions are mediated by elementary particles described by the Standard Model of particle physics.

Within the Standard Model, the strong interaction is responsible for binding subatomic particles together to form hadrons, such as protons and neutrons. The weak interaction also acts on the nucleus of atoms, helping to keep them together. The electromagnetic force, carried by the photon, creates electric and magnetic fields, which are responsible for the attraction between orbital electrons and atomic nuclei which holds atoms together. Although the electromagnetic force is far stronger than gravity, it tends to cancel itself out within large objects, so over large (astronomical) distances gravity tends to be the dominant force, and is responsible for holding together the large-scale structures in the Universe, such as planets, stars, and galaxies.

Unquote

The consensus is that all the four fundamental forces work in concert to keep things together, such as the atoms in a human body, the molecules in nature, the planets in the Solar System, and the stars in a galaxy.

A big challenge for physicists is to find a way to unite gravity in a common framework with the other three forces. In this book a fifth fundamental force of the Universe is revealed that accounts for the mentioned four forces, and that acts to bring together the micro and macro aspects of the Universe into a Final Theory of Everything (FTOE - think of the mnemonic Foot Toe).

It is theorised by some that the fourth force (gravity) is governed by 'graviton' particles. But gravitons have not been found to date and may not even exist.

“In an attempt to marry gravity with quantum theory, physicists came up with a hypothetical particle—the graviton. It is said to be a massless particle that travels at the speed of light. The graviton remains hypothetical, however, because at the moment, it’s impossible to detect” (source: Cedille de Jesus, The Edge of Physics: Do Gravitons Really Exist? Futurism.com).

"There is no direct evidence for the existence of hypothetical quantum gravity particles called gravitons" (source: Sarah Wells, A Quantum Explanation for Gravity Could Generate the Theory of Everything, Popular Mechanics, Feb. 2023).

As mentioned, the Standard Model includes the electromagnetic, strong and weak forces and all their carrier particles, and explains well how these forces act on all of the particles in matter. However, gravity is not part of the Standard Model because fitting gravity comfortably into this framework has proved to be a difficult challenge. No one has managed to make the two mathematically compatible in the context of the Standard Model.

When it comes to the minuscule scale of particles, the effect of gravity is so weak it is said to be ‘negligible’ in the sense that gravity has no significant effect on how subatomic particles behave.

Only when matter is in bulk, at the scale of the human body or of the planets for example, does the effect of gravity dominate. So the Standard Model still works well despite its reluctant exclusion of gravity among the fundamental forces.

Einsteinian relativity says that the greater the mass of an object, the greater the gravity. This is so, but not because of any Einsteinian curvature of space. It all comes down to inertia. The greater the mass, the greater the inertia. And the greater the inertia, the greater the gravity.

The concept of mass has undergone various definitions in the past. In modern physics mass is considered to be a measure of the body's inertia, meaning the resistance to acceleration (change of velocity) when a net force is applied.

So when Newton's law states that the strength of gravity between any two objects depends on its mass, this is correct. In other words, the greater the body's inertia, the greater the force of gravity. Put simply, greater mass in itself does not equal greater gravity (unless you believe in space curvature). But greater inertial mass does equal greater gravity.

To understand this better, think of two parallel roads, A and B. A very large truck on road A is accelerating faster and faster. On road B a very small light-weight golf-cart is accelerating at the same rate and speed as the very large truck. So both vehicles are keeping pace and keeping abreast with each other, and accelerating faster and faster. Imagine that each vehicle has a strong elastic band attached to the back. Now ask yourself, which of the two vehicles would be easier to pull to a stop by grabbing the elastic band with a giant hand?

The answer is that even though both vehicles are accelerating at the same rate, the big truck will be much harder to pull to a stop compared to the small golf cart. We know this empirically and intuitively, but the reason is that the bigger mass of the truck creates bigger resistance, i.e. stronger inertial gravity. The driver in the big truck will feel greater inertia and hence greater gravity compared to the driver in the small golf cart.

Gravity is caused by inertia, and inertia is caused by accelerating movement. So accelerating movement causes gravity. The lower the force of inertia, the lower the force of gravity. This is why gravity on the Moon is less than gravity on Earth: the Moon is a much smaller object with less mass compared to Earth. We will discuss this at greater length in the section on gravity.

"Inertia is the force that holds the Universe together. Literally. Without it, matter would lack the electric forces necessary to form its current arrangement and the Universe would collapse" (source: mental models, Inertia: The Force That Holds the Universe Together, fs.blog/inertia).

Any 'Theory Of Everything' needs to reconcile the four fundamental forces of the Universe into a single force that is responsible for all four forces. The outstanding question relating to the four fundamental forces is whether they are actual manifestations of just a single greater force of the Universe. If so, each of them should be able to merge with the others, and there's already evidence that they can.

For example, physicists Sheldon Glashow, Steven Weinberg and Abdus Salam won the Nobel Prize in Physics in 1979 for unifying the electromagnetic force with the weak force to form the electroweak force. Physicists working to find a so-called grand unified theory aim to unite the electroweak force with the strong force to define an electronuclear force; this has been predicted and modelled, but researchers have not yet observed an electronuclear force.

The final piece of the puzzle would then require unifying the force of gravity with the electronuclear force to develop the so-called theory of everything: a theoretical framework that could explain the entire Universe. This book provides the final piece of the puzzle.

The take home message: The four fundamental forces of the Universe are responsible for the creation of atoms and the Universe as we know it. But a fifth fundamental force is needed (and is revealed in this book) to bring the four fundamental forces together into a single *Final Theory Of Everything*.

The Accelerating Expansion of the Universe

This section shows how the Universe is expanding and why this is happening at an accelerated rate. We also look at how the Universe began, how it was shaped, and the size of the Universe.

The standard model of cosmology says that the Universe is expanding. This is confirmed by the doppler effect and the Hubble constant discussed previously. The redshift shows that galaxies are receding from Earth. To be more precise, more space is being created between things everywhere in the Universe as it expands. Think of this expansion as a balloon that is being inflated (see the following image).

The dots (galaxies) on the balloon grow apart as the balloon is inflated. Thus as the Universe expands, everything in the Universe grows apart.

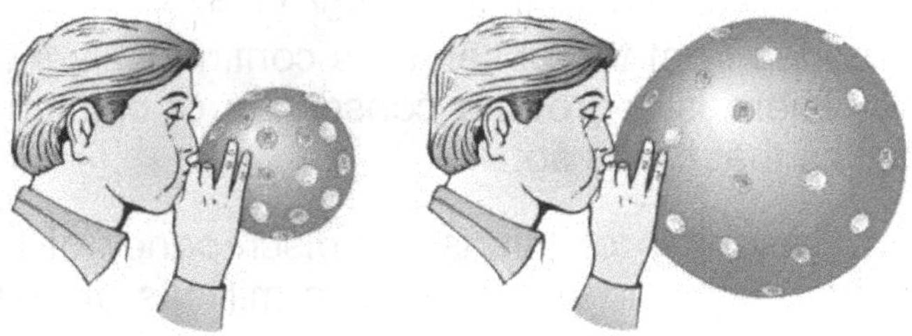

The balloon image only shows a two-dimensional expansion because we are only looking at the surface of the balloon. In reality, the expansion is three-dimensional. This means that more space is being created *around* objects (stars, galaxies, planets) from *every* direction around the objects.

Imagine raisins contained in the dough of bread. Each raisin is a very distant galaxy. As the bread is baked, the raisins separate as in the following two images:

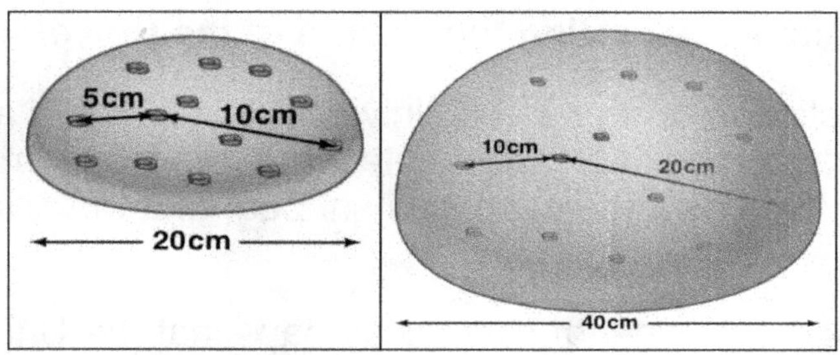

Every raisin in an expanding loaf of bread will see every other raisin moving away in all directions.

This does not imply that Earth or the Milky Way is the centre of the Universe! All galaxies will see other galaxies moving away from them in an expanding Universe unless the other galaxies are part of the same gravitationally bound group or cluster of galaxies.

Furthermore, the rate of expansion of the Universe is said to be accelerating, such that the velocity at which a distant galaxy recedes from the observer is continuously increasing with time. There is a growing consensus of opinion among cosmologists that this is so.

The main reasoning for this is the discrepancy of light that travelled from supernovae explosions millions of years ago. Such light travelled a greater distance than theorists had predicted. And this explanation, in turn, led to the conclusion that the expansion of the Universe is actually speeding up, not slowing down or remaining static.

There is some debate among cosmologists as to the rate of acceleration. Some say the acceleration is constant, others that the rate of acceleration is slowing down, and others that the rate of acceleration is speeding up. But whatever the rate of acceleration, the current consensus is that the expansion of the Universe is indeed accelerating. Even if the rate of acceleration is slowing down, it is nevertheless expanding at a faster and faster rate.

Note: In 2011 the Nobel Prize in Physics was awarded to Saul Perlmutter, Brian P. Schmidt and Adam G. Riess for the discovery of the accelerating expansion of the Universe through observations of distant supernovae.

Today, astronomers are continually observing that light from distant objects in the Universe is redshifted in whatever direction you look: all the distant galaxies are going away from us. This can only be due to the fact that the Universe is expanding. And by measuring the distance to the receding galaxies, one finds that the speed of expansion is proportional to the distance of the galaxy from Earth, i.e. that the speed of expansion is accelerating. This is called ‘Hubble’s Law’ after Edwin Hubble who was the first to discover it.

A question that arises is: *why is the Universe expanding at an accelerating rate?* Various theories have been put forward such as a repulsive gravitational interaction of antimatter, or a deviation of the gravitational laws from general relativity, or the force of dark energy. The most likely answer to this question is that the Universe has always been expanding at an accelerated rate from the moment of the Big Bang, and this is simply continuing today (this is the way the Universe came into existence).

We humans may never come to understand what triggered the creation of the Universe or how cosmic expansion works (how can new space be created from ‘nothing’). So although scientists can see that cosmic expansion is occurring, we don’t know exactly how this is happening.

The general scientific consensus is that the energy from the Big Bang drove the Universe's early expansion. But we must not assume that this initial 'propellant' of the big bang cannot still be affecting the expansion of the Universe. That the 13.8 billion light years of time is so vast that it cannot possibly still be galvanising the accelerated expansion. The Universe works on time scales that we cannot imagine, so it is entirely possible that the expansion of the Universe that we see today is due to the initial impetus of the Big Bang.

Scientists are increasingly coming round to the view that the Big Bang was not borne from a singularity (i.e. infinitely dense matter). Rather, it is thought that the Big Bang was the beginning of rapid spatial expansion and that this continues today. This spatial expansion started as a very, very small area of space (it 'sprung into existence' so to speak). And when it did so, nothing else existed at all outside of that very small space.

In other words, when the spatial expansion started nothing else existed, not even empty space. So the spatial expansion created its own space into which it could expand. This is why it is said that the Big Bang did not originate from any particular location in the Universe - the Big Bang occurred everywhere because it occurred as a very tiny point that grew and grew into the immense Universe that we see today.

The shape of the Universe

Many cosmologists have wondered about the shape of the Universe: *what is the overall shape of the Universe?* Scientists have speculated that the overall shape could be flat like a sheet of paper, or shaped like the saddle of a horse, or round with a hole in the middle like a doughnut, or even spherical. A variety of speculative theories have been put forward to explain each shape. We humans can only see a tiny fraction of the whole Universe and therefore its overall shape will always be a matter of speculation.

However, given current scientific knowledge, the most likely shape of the Universe is thought to be akin to a squashed sphere, like a round pancake:

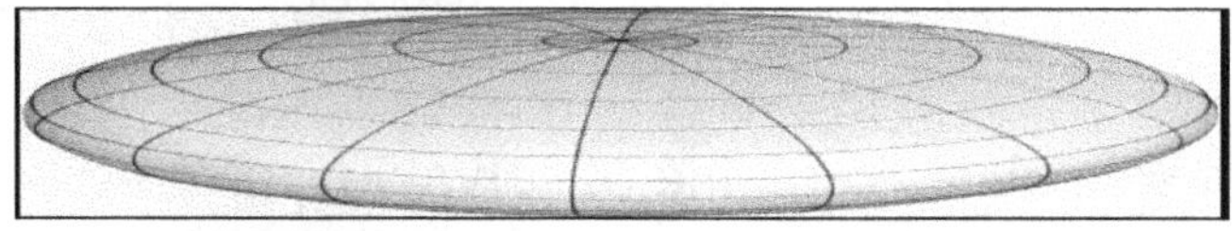

The Big Bang theory says that the Universe came into being from a single, unimaginably hot and dense point nearly 14 billion years ago. It didn't occur in an already existing space. Rather, it initiated the expansion of space itself.

The Final Theory Of Everything allows us to arrive at the pancake shape given the nature of cosmic expansion (the 'AE force') and the nature of inertial gravity (UI gravity) as postulated in this book. From the moment of the Big Bang both the AE force (Accelerating Expansion force) and UI gravity (Universal Inertial gravity) came into being. The AE force continuously creates new space into which the Universe can expand. And UI gravity makes it possible for matter to form into planets, stars and galaxies, and indeed for life to exist.

"In its first second of existence, the Universe was made up of fundamental particles, including quarks, electrons, photons, and neutrinos. Protons and neutrons then began to form. In the next few minutes, the Universe as we know it took shape" (source: PBS.org, Universe Timeline).

Since the 1980's cosmologists have theorised that the most likely shape of the Universe is like a pancake. Not completely flat like a sheet of paper with four corners, and not round with a hole in the middle like a doughnut. The flat pancake shape allows the Universe to be a closed system because the thickness of the pancake means that there is no 'edge' or 'ending' to the shape.

"Current observational evidence (WMAP, BOOMERanG, and Planck for example) imply that the observable Universe is flat to within a 0.4% margin of error of the curvature density parameter with an unknown global topology" (source: Shape of the Universe, Wikipedia.org).

One can speculate that if the observable Universe is shaped like a round pancake, then it is likely that the shape of the whole Universe is similar. The most convenient way of determining the shape of the Universe is to use the cosmic microwave background (CMB), the relic afterglow of the Big Bang. Small spatial variations in the temperature of this faint light (hot and cold spots) are produced by sound waves moving through the early Universe.

The actual size of the mentioned hot or cold spots can be computed accurately and then compared to their measured size. This is like doing a vast trigonometry measurement across the entire Universe and revealing the geometry of space.

Over the past few decades, astronomers have measured the temperature fluctuations in the CMB very accurately. The results have shown to a high degree of accuracy that the density of the Universe is such that it expands in every direction without any positive or negative curvature, in other words, the Universe is 'flat'. This flat Universe is a major component of the standard cosmological model.

"New measurements of the cosmic microwave background by the Atacama Cosmology Telescope find that the Universe is flat, with no evidence of deviation from flatness. This supports the interpretation that the deviation seen by Planck is a statistical fluctuation" (source: What Shape Is the Universe? The Quanta Newsletter, July 2020).

To continue with our speculation that the Universe is shaped like a round pancake consider the following:

1. Soon after the Big Bang subatomic particles were formed, and they in turn formed into atoms. Put very simply, subatomic particles joined together by having positive and negative charges (positive and negative attract each other). When they join, full atoms are formed. And then the atoms form into chains of atoms to make up the different elements of matter that we see today. The point here is that elongated chains of atoms tend to take the shape of flatness rather than the shape of a sphere. So by extension, as matter was formed from atoms a flat shape would predominate on the whole, leading to a flat-shaped Universe.

2. There are no known spherical galaxies, they all tend to be flat in one way or another. Also, cosmologists have never seen a group or cluster of galaxies arranged into a sphere. Cosmologists see a pattern of many galaxies extending out in all directions of the compass on a flat plane rather than in all directions as a 3-D sphere. It is therefore reasonable to think the Universe as a whole is more like a flattish pancake than a sphere.

3. When the Big Bang occurred it is speculated that it started as a growing circle on a plane rather than as a growing sphere. It is not thought that the Big Bang occurred as some kind of explosion with 'debris' spreading out in all directions. Rather, a very hot point appeared into nothingness. Nothing existed in any direction around the hot point (not even empty space). Imagine you are looking at a white sheet of paper when suddenly a tiny point of heat appears in the middle of the sheet. This tiny burning point represents an incredibly hot mix of subatomic particles at the moment of the Big Bang. At that moment there is no existence of anything behind or in front of the sheet so the burning hole can only expand in all directions on a flat plain:

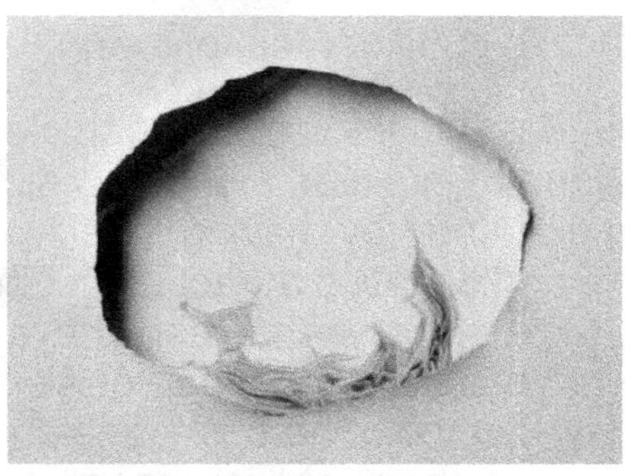

As this tiny point of very hot plasma grew into a bigger circle with little depth, subatomic particles came into existence over time, but the shape of the Universe was set from that moment onwards. Eventually the growing circle became the pancake-shape of the whole Universe.

"There isn't the slightest reason to doubt the Universe is flat. All other measurements of the CMB, like those by the Atacama Cosmology Telescope (ACT) in Chile and the Wilkinson Microwave Anisotropy Probe, are consistent with flatness. Data from other sources, most notably baryon acoustic oscillations — the imprints left on galaxies from primordial sound waves that occurred after the Big Bang — also suggest flatness" (source: Cody Cottier, What shape is the Universe? Astronomy Magazine, 23 February 2021).

The size of the Universe

The overall size of the Universe is millions of times bigger than the observable Universe. It is estimated that the observable Universe is about 13.8 billion light years in radius, say 28 billion light years in diameter in round figures. It is thought that from Earth we cannot see the Universe beyond 13.8 billion years because we cannot see beyond the age of the Universe.

However, there is some disagreement among cosmologists on this issue. Some cosmologists say that if we were to calculate how far a photon from the Big Bang could have reached Earth today, we come up with the upper limit to how far we can see in any direction: 46 billion light-years. On the other hand some astronomical and physical calculations suggest that the visible Universe is about 4% of the total Universe.

But whatever the size of the observable Universe, it is thought that the unobservable Universe is millions of times bigger (perhaps about 15 million times bigger). This makes the diameter of the whole Universe at least 23 trillion light years in diameter. Imagine that a large bed sheet represents the size of the whole Universe. Now imagine a tiny grain of sand somewhere on that bedsheet. That grain of sand represents the size of the observable Universe. The amount of the Universe that we can see (our observable Universe) is a very, very small part of the whole Universe.

The Universe is full of galaxies, stars and planets, spread around pretty evenly. It is thought that the density of all matter in the Universe is very homogeneous. So the very, very tiny fraction of the Universe that we can see is probably very representative of how the rest of the Universe must be or should be.

Given that the observable Universe is, say, 28 billion light years in diameter (based on when the Big Bang occurred about 14 billion light years ago) how can the Universe as a whole be 23 trillion light years in diameter? How did the Universe grow to 23 trillion light years in diameter over just 14 billion light years of existence?

A good question, here is the answer:

We know the age of the Universe as being roughly 13.8 billion years old. But we know much less about the overall size because we can't see it. The thing to appreciate is that the

Universe has been expanding at an accelerated rate during those 13.8 billion years. Furthermore, as the Universe as a whole grows bigger at the same percentage rate everywhere, it means the bigger the distance between objects, the greater their speed of separation. In fact, objects that are very distant from each other will be moving away from each other faster than the speed of light as the Universe expands.

So although light from the Big Bang took 13.8 billion years to reach Earth, during all that time distant objects in the Universe have been moving away from Earth faster than the speed of light. This is why the Universe as a whole is many millions of times bigger than the small amount of observable Universe that we can see.

This does not mean that galaxies are actually moving apart faster than the speed of light. It means that the expansion of space created between them is making them separate from each other at a rate that is faster than the speed of light. This is happening in the unobservable part of our Universe and this is why we cannot see things in that region.

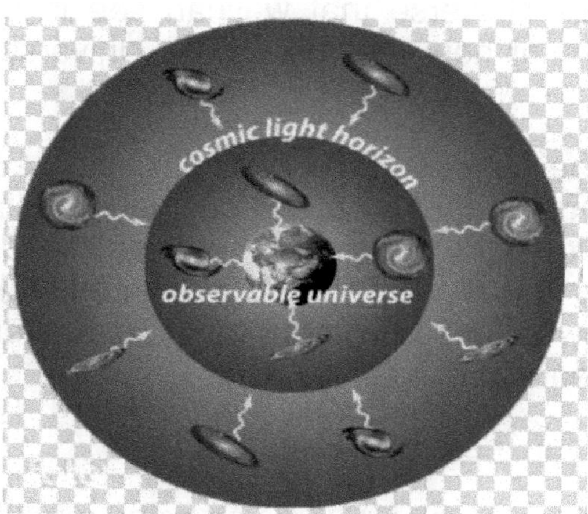

To summarise, the rest of the Universe is beyond the cosmic light horizon and we therefore cannot see it. Why not?

Because the rest of the Universe (beyond the light horizon) is so far away that any light from stars and galaxies in that region would take too long to reach Earth on human timescales; the light is too far away to reach us. In other words, the light from the unobservable Universe is indeed coming towards earth, but on human timescales it will 'never' reach us.

So we have the observable Universe and the unobservable Universe that together make up the whole Universe. It is estimated that the observable Universe is about 94 billion light years in diameter. And that the unobservable Universe is 15 million times bigger. This makes the whole Universe have a diameter of about 23 trillion light years, which is simply inconceivable for a human to grasp.

To reiterate, the unobservable Universe is so much bigger than the observable portion because while the Universe has been expanding for nearly 14 billion years it has been 'feeding' stars and galaxies into the unobservable Universe for all that time. Thus, cosmic expansion pushes stars and galaxies too far away for their light to ever reach Earth. The amount of the Universe that we can see or detect is a very, very tiny amount of the whole Universe. Nevertheless, with each passing year we see a little more of the unobservable Universe. In other words, the size of our observable Universe becomes bigger as more time passes since the Big Bang.

The overall size of the Universe is continually becoming bigger as more brand-new space is added. How is this 'brand-new space' created? What creates cosmic expansion? Another good question. Here is the answer.

What causes cosmic expansion?

When the Universe was born, nearly 14 billion years ago, it was filled with a hot plasma of particles (mostly protons, neutrons, electrons, and soon after photons). These subatomic particles were spreading out in all directions of the compass as a growing circle on a plane rather than as a growing sphere. This spreading outward movement is what causes cosmic expansion. The objects that are spreading outward need somewhere to go, i.e. space in which they can continue to spread outward.

So the movement itself (of spreading outward) is what 'compels' or 'triggers' the Universe to create new space out of what we perceive to be nothingness. And this has never stopped. Today we see distant galaxies spreading outward in all directions from our perspective. And we conclude that this spreading outward is causing cosmic expansion at an accelerated rate.

So although we can know what causes cosmic expansion, we do not know exactly how it is done. Like the mystery of the Big Bang, exactly how cosmic expansion is achieved is also likely to remain a mystery. This subject is discussed further in the section 'What came before the Big Bang?'.

The take home message: Given current knowledge, the shape of the Universe is likely to be similar to a round flattish pancake. The size of the Universe as a whole is many millions times greater than the small amount of observable Universe that we can see. The Universe is expanding at an accelerated rate, as propelled by the initial impetus of the Big Bang. We don't know if or when this rate of acceleration will slow down or stop. And indeed, it is possible that the acceleration of cosmic expansion has already started to slow down but on human time scales we cannot measure it or perceive it.

Current Theories of Gravity

This section examines the current theories of gravity and why Einstein's general relativity and his theory of gravity fail at both the macro and micro levels. The Einstein Equivalence Principle (EEP) is also discussed.

Theories of the Universe depend on an accurate understanding of gravity. The standard model considers that gravity is the only force in physics that affects matter on very large scales. But gravity alone can't explain certain astronomical observations. For example, if we measure the speed of stars on the outskirts of a galaxy, it is thought that they're moving too fast to remain in orbit if the only thing holding them back is the gravitational pull of the visible galaxy. Similarly, clusters of galaxies appear to be held together by a stronger force of gravity than can be accounted for by any current explanations.

In mainstream physics, there are currently four main proposed solutions: 1. The SMC model, 2. The MOND model, 3. The M-theory model, and 4. The general theory of relativity model.

The SMC model

This is the Standard Model of Cosmology. It postulates that the Universe was created in the 'Big Bang' from a singularity of pure energy, and is now composed of about 5% ordinary matter, 27% dark matter, and 68% dark energy. The SMC model assumes a cosmological constant (an all-pervasive ether made of dark matter). It also assumes that cosmic expansion is propelled by dark energy. The SMC model is wedded to Einsteinian relativity by requiring a cosmological constant that makes it possible for general relativity and spacetime gravity to work. Both dark energy and dark matter are simply names describing unknown entities. And dark matter is the modern-day equivalent of an all-pervading ether.

Doubts about the SMC model are gradually surfacing. Here is an example:

"The current SMC has numerous dubious assumptions arising from the dubious assumption that general relativity is valid for all distances within the expanding Universe and not just for the very small distances of the Solar System. Also, the SMC assumes the existence of both dark matter and dark energy. The dark matter hypothesis is very dubious, since to date no dark matter has been definitely detected and the nature of dark matter remains unknown. And regarding dark energy it is a hypothetical form of energy that pervades the whole of space and causes the expansion of the Universe to accelerate at large cosmological distances. Currently there exists no accepted physical theory of dark energy, suggesting that the existence of such energy is a dubious assumption of the SMC" (abridged extract from: Brian Albert Robson, Redefining Standard Model Cosmology, IntechOpen, 10.5772/intechopen.85605, April 2019).

It is concluded that the SMC is baseless for several reasons, but mainly because it requires the existence of an all-pervading ether. Without such an ether, gravity as postulated by general relativity cannot exist. Einstein himself said that when objects are close enough to each other they trigger gravity between them, and the only way this can happen is through the medium of an ether, otherwise such objects would not perceive each other's existence or be affected by each other's gravity. Contemporary scientists who support the SMC argue that yes indeed there must be an ether and that it takes the form of dark matter. But dark matter remains completely unproven and undetected.

The MOND model

The maverick alternative is that the SMC model of gravity is wrong, and should be replaced by something called Modified Newtonian Dynamics (MOND); this was proposed by scientists in 2002 in the journal Annual Review of Astronomy and Astrophysics.

The two options, SMC and MOND are equally consistent with some macro observations of the heavens, but are yet to be proven sufficiently to be credible. However, these two options do nothing to explain micro-observations of quantum gravity.

MOND is one of several alternative theories or ad hoc ideas relating to gravity. In fact, MOND is a simplistic formula that violates even basic conservation laws. The formula fits spiral galaxy rotation curves reasonably well, consistent with the empirical Tully-Fisher law that relates galaxy masses and rotational velocities. But MOND fails for just about everything else, including low density globular clusters, dwarf galaxies, clusters of galaxies, and many other cosmological observations.

The M-theory model

M-theory presents an idea about the basic substance of the Universe. But to date science has produced no experimental evidence to support the conclusion that M-theory is a description of the real world. Although a complete mathematical formulation of M-theory is not known, the general approach is to offer a universal "Theory of Everything" that unifies gravity with other forces such as electromagnetism. M-theory aims to unify quantum mechanics with general relativity's gravitational force in a mathematically consistent way. In comparison, other theories such as loop quantum gravity are considered by physicists and researchers/students to be less elegant, because they posit gravity to be completely different from forces such as the electromagnetic force.

In physics M-theory attempts to unify several versions of superstring theory by showing that such theories are all related. Aspects of general relativity, quantum field theory, superstring theory, and multidimensional theory are all marshalled together to try and show a theory of everything. To date there is no complete formulation of M-theory and there is no experimental or demonstrable evidence of any kind to support M-theory.

The General Theory of Relativity model

When Einstein was asked to explain the fundamental nature of gravity he postulated two fundamentally different kinds of gravity in his theories of relativity, existing side by side:

1. **Einsteinian gravitational mass (referred to as 'spacetime gravity' in this book).** According to relativity, gravity is caused by the mass of an object. The mass is said to interact with spacetime to create a curvature of space around the mass. When an object approaches the said mass it curves around the mass (or falls into it) by being obliged to follow the curvature of space. Put another way: a massive object generates a gravitational field by warping the geometry of the surrounding spacetime.

For example, when you fall off a chair and hit the ground, there is no force of attraction (no pulling or pushing force); you hit the ground by being obliged to follow a curved line (a curvature of space) that leads you from the chair to the ground.

2. **Inertial mass.** According to Einsteinian relativity, gravity

can also be caused by the inertial mass of an object and has nothing to do with the curvature of space. Inertial mass occurs by creating inertial resistance to the acceleration of a body or mass when responding to some type of force.

Note: In this book we treat the word 'inertia' as a noun and the word 'inertial' as an adjective.

Is inertia a 'force'? In Physics, inertial force is defined as the push or pull on an object with mass to cause it to change its velocity or direction. Force is an external agent capable of changing a body's state of rest or motion. Clearly, inertia is not an 'external agent'; it is a passive property (the resistance to a change in movement). So strictly speaking inertia is not a force. But in this book we use the term 'inertial force' loosely to refer to the effect or strength of inertia.

To clarify further, inertia is a measure of mass, momentum is a measure of mass moving at a constant velocity, and inertial force is a measure of mass changing velocity. So mass is simply the amount of matter in an object. And 'inertial mass' (same as 'inertial force') is the force of inertia manifested by the mass.

Inertial gravity is easy to understand. For example, when you accelerate while sitting in a car your human body (i.e. the mass of your body) creates inertial resistance to the force of the car accelerating. Inertia means an object will continue its current motion until some force causes its speed or direction to change.

Thus when the car speeds up, the object (human body) undergoes a change in motion from the force exerted by the car. The car pushes against your body. This change in motion (pushing you against your seat) is known as inertia.

For the sake of completeness it is mentioned that some

relativists argue that space is not curved, but compressed. The commonly depicted gravitational well resembling a trampoline is said to be a misrepresentation. Rather, spacetime gravity is said to work along curved geodesic lines to form a lattice.

In this compressed lattice, time maintains the same pace locally relative to the local space-time metric. But from outside the lattice an observer sees a longer time for the light to travel a given distance as measured by the outsider's standard. This means space is compressed, the speed of light slows down relative to the outside standard, and time passes slower. This contradicts the Einsteinian postulate that light always travels at the same constant speed for all observers regardless of their location or motion.

False equivalence principle

Now we come to Einstein's so-called 'equivalence principle'. Here is a description from Equivalence Principle, Wikipedia.org:

"In the theory of general relativity, the equivalence principle is the equivalence of gravitational and inertial mass, and Albert Einstein's observation that the gravitational 'force' as experienced locally while standing on a massive body (such as the Earth) is the same as the pseudo-force experienced by an observer in a non-inertial (accelerated) frame of reference".

Put another way, Einstein's equivalence principle (abbreviated to EEP) is said to be the equivalence of spacetime gravity and inertial mass. Spacetime gravity caused by spacetime curvature is said to be equivalent to inertial mass caused by acceleration. They are said to be equivalent in the sense of having the same gravitational effect and the same mathematical description.

EEP is considered to be the cornerstone of the general theory of relativity. If EEP is false, Einstein's general theory totally falls apart. Here is why EEP is indeed false.

Einsteinian general relativity says that gravity is caused by a curvature of space that makes objects move together. This space curvature is caused by the mass of an object. A good analogy is to think of a trampoline. The bigger the mass of an object, the bigger the 'dent' of the object on the trampoline. And the bigger the dent, the bigger the curvature of space around the object (let's call this 'object A').

If another object's space curvature overlaps with object A, the two objects will move towards each other by following a curvature of space (the curved lines on the trampoline). In the following image the three objects are not close enough to fall prey to each other's spacetime gravity, so they stay apart:

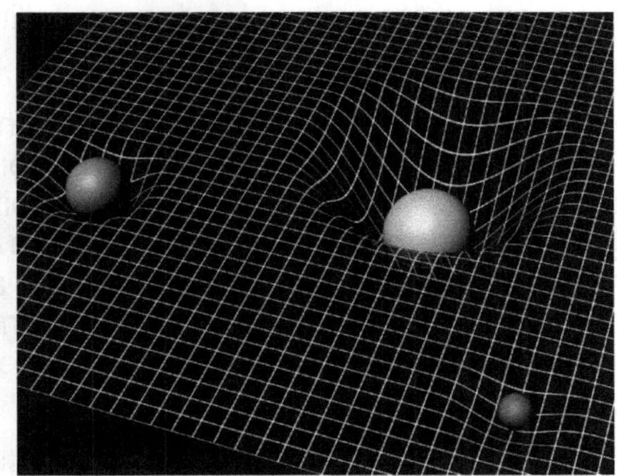

So we have the following scenario. Spacetime gravity is triggered when the space curvatures of objects overlap, causing objects to move together. Spacetime gravity is not triggered when the space curvatures of objects do not overlap (as in the above image). Einstein was clear in saying that gravity caused by space curvature is not a pushing or pulling force at all, and indeed that gravity is not any kind of force at all. Quite simply, in relativity gravity is caused by the curvature of space making objects follow curved lines towards each other.

This being the case here is a question for you dear reader: given that all the planets in the solar system are well within the Sun's curvature of space, why don't the planets follow the lines of space curvature towards the Sun? Why doesn't the Earth fall into the Sun?

Here is the answer given by relativity advocates: The planets are constantly falling towards the Sun, but due to their high velocity they always "miss it". If planets had less velocity than needed, they would indeed fall into the Sun. If they had more speed they would escape from the Sun and leave the solar system. They stay in orbit because their velocity is exactly the same as that required for neither leaving nor falling into the Sun.

The aforementioned explanation is correct. In contemporary physics it is well-established that the angular momentum of planets around stars is what keeps planets in orbits, but this has nothing to do with the relativity of spacetime curvature. The movement of Earth creates angular velocity and hence angular momentum, and inertia arising from angular momentum stops the Earth from falling into the Sun and ensures the Earth continues in its orbit.

So how exactly does spacetime curvature explain that the Earth does not fall into the Sun in spite of being well within the Sun's dent on the trampoline so to speak. What is stopping the Earth from following the lines of space curvature towards the Sun?

The general theory of relativity emphatically stipulates that the gravity of spacetime curvature does not involve any kind of force and that the only thing that makes objects go towards each other is space curvature. And once an object is caught in those curved lines on the trampoline of spacetime curvature you cannot escape. Why not? Because it is postulated that any object following spacetime curvature towards another object 'thinks' it is going straight.

In other words, as an object follows this curvature, then from the object's point of view it is not following a curved path - it is going straight as if nothing has happened. And furthermore, things like angular momentum and inertia don't exist in spacetime gravity because such gravity is defined as not being any kind of pulling or pushing force. Logically then, spacetime gravity should be sending planet Earth straight into the heart of the Sun. In reality the forces of inertial gravity and angular momentum prevent this on human time scales. Hence, EEP is spurious because the falsehood of spacetime gravity cannot be equal to the reality of inertial gravity.

It is claimed that tests of general relativity have reached high precision. Such tests refer to light deflection, the Shapiro time delay, the perihelion precession of Mercury, the Nordtvedt effect in lunar motion and gravitational wave damping. However none of these tests have been able to validate general relativity.

For example the Shapiro time delay refers to a time delay caused to light's journey when it passes a massive object and then continues on to Earth. The time delay is said to be caused by 'spacetime dilation' and the curvature of space that makes light bend around a massive object before such light continues its journey onto Earth and into the eyes of observers.

The following image shows light following a longer path 'b' (according to the Shapiro time delay) compared to path 'a':

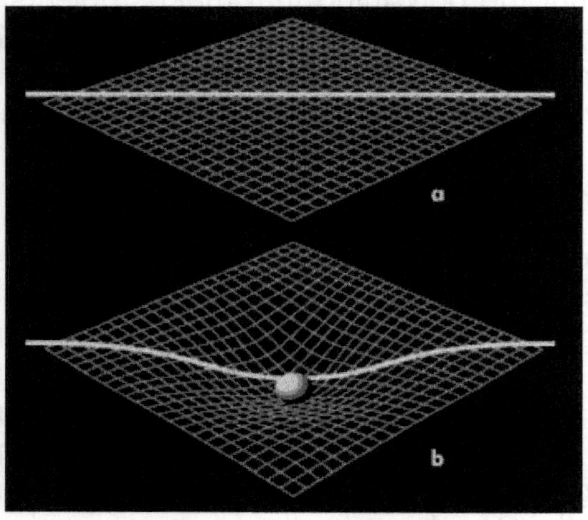

In reality, light can never bend, it can only ever travel in straight lines. So light can never be delayed by having to bend around a massive object in its journey to Earth. If the light that reaches Earth has passed a massive object, it simply means that such light continued in a straight line as if the massive object was not present.

An experiment often quoted by relativists refers to 'Gravity Probe B' (GP-B). This experiment was launched in 2004. It was a satellite-based gyroscope experiment funded by Nasa with the objective of proving the veracity of general relativity.

At the time, GP-B made headline news saying that Einstein's general theory of relativity had been proven to be correct. The experiment was deactivated in 2010 amid comments that the results were too unreliable to make it worthwhile continuing. Two main issues arose: solar flares and Earth's movement.

The regular occurrence of solar flares made the GP-B results too unreliable. NASA recommended against extending the experiment beyond 2008, warning that the effect of solar flares *"is so large that any effect ultimately detected by this experiment will have to overcome considerable (and in our opinion, well justified) scepticism in the scientific community"* (source: Hecht, J, Gravity Probe B scores 'F' in NASA review, New Scientist. May 2008).

The other issue, hardly commented on in the press, is Earth's movement. The Earth follows a complicated dance through space with many wobbles and movements that cannot be anticipated accurately and accounted for to the degree required by the GP-B gyroscope experiment. Robert O'Connell, a theorist at Louisiana State University, commented that *"This $760 million experiment was government money and to my mind it was misspent and poorly managed"*.

If we are to accept the reality of Einsteinian gravitation mass it means every physical object (anything with mass) will create a curvature of space around itself, from the pencil on your desk to whales in the ocean. Every rock and insect will be experiencing their own separate gravity field based on how their mass creates curved space around them.

And these trillions and trillions of little gravity fields will all be overlapping with Earth's gravity field in a constantly changing panorama of interweaving and overlapping gravity fields. Can this really be so?

If it is argued that the curvature of space only occurs around large objects such as planets and stars, then such an argument has to be baseless. When does spacetime gravity kick in? Exactly how big (or small) must the object be in its mass for space to curve around and cause gravity? If there is a cut-off point, does that mean small objects below the cut-off point don't have gravity?

"Unfortunately, Einstein left some basic questions unanswered: on what scales does matter tell space how to curve? What is the largest object that moves as an individual particle in response? And what is the correct picture on other scales? These issues are conveniently avoided by Einstein and Friedmann" (source: David Wiltshire, Professor of Theoretical Physics, University of Canterbury, Can we ditch dark energy by better understanding general relativity? June 2017).

"The false concept of the equality of gravitational mass [spacetime gravity] *and inertial mass was introduced into physics by Albert Einstein around 1916, because he needed the weight, heaviness, gravitational acceleration, and inertia of any mass to be equal or even identical in order to justify his ad hoc General Theory of Relativity"* (source: Justin M Jacobs, The Relativity of Gravity, work in hand treatise, 2023).

The point here is that any theory of gravity must be applicable to all types of masses in the Universe. Unfortunately the Einsteinian theory of 'gravitation mass' does not work for small objects such as subatomic particles (the mathematics falls apart). Yet it is known that atoms have gravity. If they didn't have gravity, atoms would not 'stick together' to form objects containing millions and trillions of atoms. Even the protons and electrons inside atoms are known to have gravity, however little.

It may be argued that Einsteinian gravitation works for very big objects, and that there must be another type of gravity that works for very small objects. This would imply one kind of gravity for the big and another fundamentally different kind of gravity for the small - a very unlikely scenario.

Furthermore, the latest research is showing that Einstein's Theory of Gravity fails at both the macro level (large cosmological level) and the micro level (subatomic particle level).

The standard theory of the cosmos in today's world describes all the visible matter and energy in the Universe, and shows how this visible matter has evolved according to Albert Einstein's theory of gravity. Einstein postulated that the Universe was static and from the 1930s until the late 1990s, most physicists agreed with Einstein's choice of setting the cosmological constant to zero.

But Einstein's cosmological constant was abandoned after Hubble's confirmation that the Universe was expanding, and indeed accelerating in its expansion. Einstein regretted modifying his elegant theory and viewed his cosmological constant as his 'greatest mistake'.

Other cracks are appearing in Einstein's Theory of Gravity on a macro level. Einstein suggested that the fabric of space is smooth and continuous, like a stage that remains in place whether actors are treading its boards or not — even if there were no stars or planets dancing around, spacetime would still be there. However, physicists Laurent Freidel, Robert Leigh, and Djordje Minic think that this picture is holding us back. They believe spacetime doesn't exist independently of the objects in it. That spacetime is defined by the way objects interact.

The gravity of spacetime curvature is full of contradictions. For example, when a rocket leaves Earth's gravity it means it leaves Earth's curvature of space. This means that if the force of the rocket is strong enough it can break away from space curvature. So at a certain point the pulling force of space curvature is not strong enough to stop the rocket from escaping. But space curvature has no pulling force!

In reality, it is well-established in physics that all objects in the

Universe have inertia by virtue of their movement, mass and momentum. Such inertia is not some kind of force that reaches out to push or pull. Here, any reference to the 'force' of inertia is used loosely to refer to the effect of inertia caused by movement, mass and momentum; this force of inertia is well understood by science.

Indeed, the mathematics of the force of inertia are well-established. The mathematics of the equivalence principle is simply a set of theoretical spacetime-curvature-mathematics contrived and manipulated to be equal to the mathematics of inertia. By making this link between the theoretical, fictitious nature of spacetime curvature and the real world of inertia, it greatly helped give credibility to the theory of spacetime gravity as postulated in Einstein's general theory of relativity.

In short, spacetime gravity is spurious and non-existent, and this means there can be no equivalence principle between non-existent spacetime gravity and fully existing inertia. Scientists readily admit they don't know why spacetime gravity and inertial mass are equal mathematically, but now we know: they are not equal.

Dark Energy

The largest galaxy survey ever made to date (the DES project, darkenergysuyrvey.org) has revealed a discrepancy with Einstein's general theory of relativity. Data from this survey published in 2018 (and still ongoing) concluded that *"our understanding of how giant structures grow and evolve over cosmic time — which rests on our understanding of gravity through Einstein's general theory of relativity — might be wrong".*

Yet another crack in Einstein's Theory of Gravity relates to 'empty space'. Einstein said that empty space can possess its own energy. Because this energy is a property of space itself, it would not be diluted as space expands. As more space comes into existence as a result of the expansion effect, more of this energy-of-space would appear. As a result, this form of energy would cause the Universe to expand faster and faster.

Today, cosmologists generally do not accept Einstein's theory of 'empty space', and instead talk about a mysterious and hypothetical form of energy called 'dark energy' that is used to explain the accelerating expansion of our Universe. However, the existence of dark energy has never been shown to exist, and research is increasingly showing this.

"A new model raises doubt about the composition of 70% of our Universe – dark energy may simply not exist" (source: University of Copenhagen - faculty of science, March 31, 2021).

"Dark energy is the biggest mystery in cosmology, but it may not exist at all" (source: Professor Subir Sarkar, head of the particle theory group at the University of Oxford, UK).

"Scientists have failed to detect dark energy in a laboratory. Also, experiments to detect dark energy through its interaction with baryons in the cosmic microwave background has also led to a negative result" (source: abridged extract, Cosmological Constant, Wikipedia.org).

"Dark energy and dark matter are theoretical inventions that explain observations we cannot otherwise understand. Einstein mistakenly wanted to exactly balance the self attraction of matter by anti-gravity on the largest scales. He could not imagine that the Universe had a beginning and did not want it to change in time" (source: David Wiltshire, Professor of Theoretical Physics, University of Canterbury, Can we ditch dark energy by better understanding general relativity? June 2017).

The argument for the existence of dark energy goes like this: To have an accelerated expansion of the Universe we need energy pumped into the system. And we don't see the source of this energy that is causing this accelerated expansion. Dark Energy is a hypothesis to explain this accelerated expansion.

The argument against the existence of dark energy goes like this (two points):

1. The Accelerated Expansion force (AE force) came into being the moment the Big Bang occurred. This AE force gives the Universe the ability to create new space into which it can expand. We do not understand exactly how the AE force works, i.e. exactly how the Universe can create new space out of 'nothingness'. Nevertheless, contemporary cosmology is unanimous in thinking that the Universe as a whole is indeed becoming bigger as evidenced by the redshift, the Hubble Constant and other cosmic phenomena.

2. The expansion of the Universe is at a constant accelerated rate. It is accelerated because the Universe as a whole is becoming bigger and therefore the rate of growth must be at an accelerated rate. For example, on Monday the diameter of object X as a whole grows 2% bigger. On Tuesday it grows another 2% bigger. Now compare the diameters of object X on Monday evening and Tuesday evening. You will find that on Monday the diameter grew by 10 cm, but on Tuesday it grew by 12cm even though on Tuesday the overall growth rate was the same 2%. Hence, each day the overall growth rate is at an accelerated rate. The accelerated rate comes from applying the same percentage rate of expansion growth, but to an ever-bigger size.

No dark energy is needed to account for the accelerated growth rate of the Universe. But we do need a new model that explains how the Universe works without recourse to hypothetical dark matter or dark energy. This subject is fully covered in the book.

Gravitational Waves

According to Wikipedia.org, gravitational waves are thought to be generated by the accelerated masses of an orbital binary system, and propagate as waves of gravity radiation. This is a far cry from explaining the nature of gravity that we experience in our daily life. The existence of gravity waves is yet to be proven.

Einstein was mostly ambivalent about the nature of spacetime, unwilling to say clearly whether spacetime, which he described as a curvature of space, was or was not composed of some kind of particles or waves. He faced a contradiction: if physical mass creates a curvature of space, it means that a curvature of space emanates from something physical. So something physical must exist for space curvature (and hence gravity) to exist.

Yet at the same time Einstein insisted that space curvature was absolute, that it existed independently of any kind of matter. His famous phrase *'spacetime is like a stage that remains in place whether actors are treading its boards or not'* comes to mind.

Albert Einstein thought long and hard about the possible existence of gravitational waves, but he remained sceptical about them most of his life and then 'proved' their non-existence in his general theory of relativity.

However, some cosmologists argue that Einstein did in fact predict the existence of gravitational waves in his general theory of relativity. This is not so. Einstein predicted the existence of gravity as a curvature of space - not as waves, and not as any kind of cosmic ripples travelling at the speed of light.

Attempts to detect gravitational particles or waves have been made, but to no avail. A preliminary announcement about finding gravitational waves was made in 2014, but was quickly retracted, after astronomers found that the signal detected could be explained by dust in the Milky Way (source: Our expanding Universe: Age, history & other facts, space.com).

You may have heard that the LIGO experiment in 2015 discovered gravitational waves. It was claimed their results showed the existence of '*ripples or undulations in spacetime*' that the scientists called 'gravitational waves', but it came to nothing. LIGO may well discover some kind of cosmic phenomena in the future, but proving that such phenomena are waves of gravity is another matter entirely.

"We believe that LIGO has failed to make a convincing case for the detection of any gravitational wave event; it was all an illusion" (source: Andrew Jackson, Emeritus, Theoretical Particle Physics and Cosmology, Niels Bohr International Academy).

Any claim that gravitational waves exist would contradict Einsteinian relativity. Towards the end of his life Einstein was clear that gravity is a distortion of space caused by the mere presence of matter or energy, such as a planet or star. In other words, Einstein was saying that gravity makes the movement of two or more objects go towards each other, and that the mere presence of matter is enough to create gravity.

Einstein vehemently argued that gravity is not any kind of force. He argued that objects attract each other because their presence causes the objects to veer towards each other as a result of 'curving' space. This is what his general theory of relativity is all about.

The question that arises is the following: *how exactly does spacetime make space become curved so as to make two or more objects come together due to gravity?* Scientists have wrestled with this question ever since Einstein published his general theory of relativity; they have tried to understand the true nature of spacetime curvature, but without success to date.

Scientists in this area of study have tried to answer the question by saying '*spacetime fuses the three dimensions of space and the one dimension of time into a single four-dimensional manifold called spacetime. This spacetime is what curves the space between objects when they get close enough, making them go towards each other*'.

But in saying this they readily admit they do not understand how spacetime creates a curvature in space called gravity. This lack of understanding generates alternative theories such as a belief that gravity is some kind of absolute phenomenon of force or that gravitational waves or gravitons must exist. None of these ideas have been verified. Even the theory that gravity exerts a force of attraction is baseless and unproven, and on this point Einstein fully agreed.

Gravitational constant

Staying on the topic of the current theories of gravity, scientists have not yet managed to measure and establish a so-called 'gravity constant' (also referred to as a gravitational constant or cosmological constant) although efforts continue.

"A gravitational constant for the Universe cannot be established mathematically, only empirically. It is difficult to measure with high accuracy. This is because the gravitational force is an extremely weak force as compared to other fundamental forces at the laboratory scale" (source:

Wikipedia, Gravitational Constant).

Measuring gravity in modern physics is very much a hit and miss affair, and whatever the measurement it can only ever be a rough approximation. Gravity is measured by the acceleration that it gives to freely falling objects. At Earth's surface the acceleration of gravity is about 9.8 metres (32 feet) per second per second. Thus, for every second an object is in free fall, its speed increases by about 9.8 metres per second.

This yardstick of 9.8 metres is then used to compare gravity in other places. For example it is calculated that the gravity on the moon is about 16.6 % that of Earth. So mathematically, the force of gravity is calculated as the rate of acceleration caused by a free-falling object. And this rate of acceleration is determined by the amount of mass that is receiving the free-falling object.

Clearly, gravity can vary from place to place on the surface of Earth depending on the mass at any particular spot. Geophysical conditions below ground, the latitude location, the wobble of Earth's rotation, distance to Earth's centre and other factors can all affect the amount of mass from location to location measured at the surface. It is estimated that the strength of gravity at Earth's surface can vary by as much as 1 percent. Because of this a universal mathematical constant for the value of gravity cannot be established and 9.8 metres per second is used as an agreed rough yardstick.

Gravimetry instruments have been developed to measure gravity more accurately: *"For a small body, the gravitational effects predicted by general relativity are indistinguishable from the effects of acceleration by the equivalence principle. Thus, gravimeters can be regarded as special-purpose accelerometers. Many weighing scales may be regarded as simple gravimeters. Researchers today use more sophisticated gravimeters when precise measurements are needed"* (source: Wikipedia.org, Gravimetry).

As mentioned, when the force of gravity is measured on the surface of the Earth, it varies from place to place. This is because the planet is not perfectly spherical or uniformly dense. In addition, gravity is weaker at the equator due to centrifugal forces produced by the planet's rotation.

Einstein attempted to introduce a gravity constant into his scientific revelations, but later withdrew it. Equally he also withdrew his idea of a static Universe, and grudgingly accepted the idea of a dynamic Universe that was continually expanding.

Simply put, a gravity constant could not work the way Einstein intended because it would not allow the kind of static Universe that he was trying to match. That error arose in part because once again Einstein used the wrong coordinate frame for his calculations. But also his concept was wrong from a physical perspective. Although it is possible to briefly balance the gravitational attraction of matter with the repulsion from a cosmological constant, the smallest perturbation would produce runaway expansion or collapse. With or without the cosmological constant, the Universe must be dynamic.

The take home message: Current theories of gravity fail to provide a theory of gravity that works for both big things (planets and stars) and small things (subatomic particles). Hitherto the underlying nature of gravity has been unknown, and any measurement of gravity can only give an approximation. A 'gravity constant' in the form of an ether is considered baseless.

The Fifth Fundamental Force

In this section a fifth fundamental force of nature is revealed, called the AE force, that brings together the gravity of the big and the small. The underlying nature of gravity is also discussed, setting the scene for the *Final Theory Of Everything.*

The AE force (Accelerating Expansion force of the Universe) creates new space around everything that exists; it is doing this continually at an accelerated rate. In doing so the AE force makes all objects (or groups of objects) move apart at an accelerated rate, in all directions. This accelerated moving apart in all directions creates inertia in all objects in all directions. This all-encompassing inertia around all objects is what we know as gravity.

The AE force came into being at the same moment that the Big Bang created the Universe. When the Big Bank occurred it needed space in which to occur. So as soon as the Big Bang occurred the Universe started to expand at an accelerating rate. This accelerating expansion brought into being the AE force and with it the gravity that we are all familiar with, referred to in this book as UI gravity (Universal Inertial gravity).

Everything that exists has gravity, from the smallest subatomic particles to the biggest objects in the Universe. Here is a thought experiment to explain this further:

Imagine Tony on planet Earth looks through a telescope at a distant galaxy and sees that the distant galaxy is moving away from Earth, as corroborated by the redshift. In that distant galaxy an alien is doing the same: he is looking through his telescope and sees that the Milky Way galaxy is moving away from him because the Milky Way is redshifted in his telescope.

So both Tony and the distant alien are moving away from each other at an accelerating rate. This accelerated moving away from each other creates inertia around both Tony's planet and the alien's planet. It can be compared to two stationary cars, parked back-to-back. They both accelerate away from each other in opposite directions; as they do so both drivers will feel inertia pushing them back into their seats from the car acceleration.

So although Tony on Earth and the alien in a distant galaxy are not accelerating away from each other in cars, they are nevertheless accelerating away from each other due to the new space being created between them, arising from the AE force.

As the AE force creates new space around objects (by separating them like the raisins being baked in the dough of bread), the objects move away from each other in all directions. So everyone on Earth will feel the gravity of inertia, not just Tony looking through his telescope. And every alien on the distant planet will also feel gravity, not just the alien looking at Tony through his telescope.

In fact, everything that exists in the Universe is continually being imbued with gravity; every galaxy, every planet, every grain of sand, every subatomic particle is imbued with gravity.

You may be asking yourself the following: *Many things are not separating and are not experiencing new space between them, such as the atoms in my body or the planets in the Solar System, or the grains of sand on a beach, or the things in my refrigerator. So how can it be said that the AE force is putting new space around everything that exists so as to imbue them with gravity?*

The answer is that the AE force is indeed continually doing this. It is putting more space around everything that exists thus making them move apart at an accelerated rate in all directions. And in so doing, giving them inertial gravity from the moving apart movement.

As an example, let's take the things in your refrigerator. Everything in the refrigerator is moving away from distant objects in the Universe at an accelerating rate, in all directions. This gives gravity to everything in the refrigerator.

Since everything in the refrigerator has gravity, and since the refrigerator and its contents are close to Earth, both objects (the refrigerator and Earth) are caught in each other's gravity, so they stay put. As planet Earth is the biggest mass, its gravity will dominate all the smaller things on its surface, and that is why the refrigerator and its contents stay put. That is why the grains of sand on a beach or the atoms in your body stay together. That is why the stars and planets in the Milky Way stay together to form a galaxy.

On the one hand you have the AE force that pushes things to separate (moving things apart) to create gravity through inertia. And on the other hand you have the same AE force that pushes things together (coalescing force) if they are close enough to be affected by each other's gravity.

When objects are sufficiently distant from each other to not be affected by each other's gravity, they move apart. Inertial gravity is always a pushing force, whether pushing things apart or pushing things together

So we have two possible scenarios in the context of anything and everything moving in the Universe:

Scenario A: Coalescing. Objects coalesce (come together) if they are close enough to be affected by each other's gravity. Orbits are a form of coalescing.

Scenario B: Separating. Objects separate (move apart) if they are not close enough to be affected by the gravity of other objects.

Note: In the next section 'The Coalescing Force' we explain more fully how the AE force pushes things together.

In above scenario A, all coalescing objects are also simultaneously separating from distant objects in all directions. In scenario B nearly all separating objects will simultaneously be coalescing with nearby objects.

Here is an abridged extract from Wikipedia.org, Expansion of the Universe:

"Expansion [of the Universe] *is a key feature of Big Bang cosmology. It is modelled mathematically with the Friedmann–Lemaître–Robertson–Walker metric and is a generic property of the Universe we inhabit. However, the model is valid only on large scales (roughly the scale of galaxy clusters and above), because gravity binds matter together strongly enough that metric expansion cannot be observed on smaller scales. As such, the only galaxies receding from one another as a result of metric expansion are those separated by cosmologically relevant large scales".*

Put simply, the above quote is saying that in the 'Standard Model' of modern cosmology the Universe is expanding at a uniformly accelerating rate, and that gravity keeps things together if they are close enough to be affected by each other's gravity. But the gravity of subatomic particles and the underlying nature of gravity are not explained.

So the question is: *what is gravity?* contemporary science tells us that gravity can now be understood and calculated as a pseudo force. In science a pseudo force is a force of inertia that acts on a mass. A pseudo force is also referred to as a fictitious force in the sense that such force does not exist on its own as an absolute free agent. Gravity arises from inertia, and inertia arises from a mass when such mass is forced to change its speed or acceleration.

Let us consider a lone star that is not near any galaxies and is not being affected by the gravity of any other object (a rare, but possible occurrence). This lone star will always be moving away from all distant objects in all directions. As a result it will have its own gravity by virtue of being separated from distant objects in all directions, and it will not be coalescing with other objects. But one day perhaps, the lone star may come close enough to another object and coalesce.

Within galaxies small things are attracted to bigger things and so on up the food chain, ending up with giant stars that end up orbiting the galaxy with many other stars and objects. They are orbiting a common centre of mass, which may or may not be a giant black hole.

It is thought that the Milky Way has a giant black hole, close to 4 million solar masses. And the greater the mass the greater the inertial gravity. It is not known whether the black hole alone accounts for the orbiting nature of the galaxy. It is more likely that the AE force is acting through inertial gravity to gradually bring objects together or into orbits, over many years.

There is no other type of movement in the Universe. Things are either coalescing or moving apart. It can rightly be argued that things also do both. For example, the Earth is coalescing with other things in the Solar System and the Milky Way. But at the same time, the Earth is moving away from distant galaxies, because the whole Milky Way is moving away from distant galaxies. So it depends on your perspective.

Standing Man Analogy

To better explain the nature of gravity we will imagine a man in England named John standing normally on the surface of our planet Earth. Imagine that as the AE force moves the Earth rapidly upwards from where John is standing, John will be pushed to the surface of the ground (so as to stay standing) as a result of inertia. In other words, as the planet moves upwards everything on the surface (that is facing upwards) will be rooted to the ground as a result of inertia.

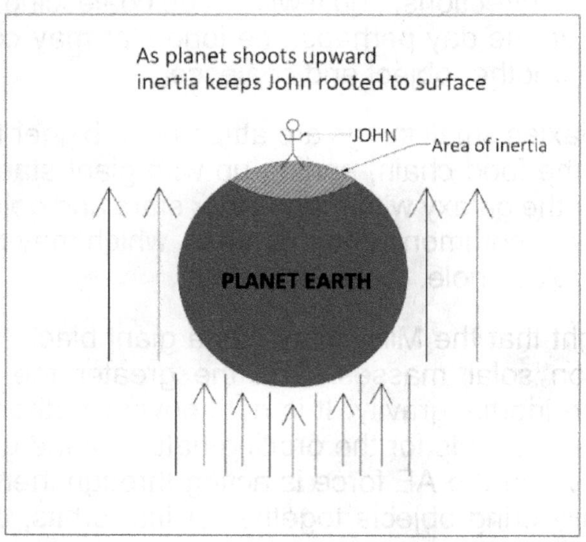

What about Paul in Australia who is also standing upright on the other side of the world? Paul will also be subject to the same force of inertia as John in England. This is so because when the AE force creates more space around objects it does so from *all directions* around objects. Thus, the AE force exerts a pushing force against all surface areas.

The AE force does not move anything in a linear sense, from A to B. It moves objects apart from each other by creating more space *around* and between objects. In doing so the force of inertia is applied around the object, everywhere. This force of inertia is what we know as gravity. It is inertial gravity and it has a pushing effect - we are continually being pushed to the surface of Earth.

This concept may be difficult to visualise because we humans cannot understand intuitively how the Universe is expanding, and in so doing, how it is creating more space between objects *'from nothing'*. In other words, we cannot conceive how the Universe is continually creating completely new additional space by making the whole Universe bigger (remember the baking bread analogy). However, the standard model in today's cosmology does indeed accept that this is what is happening, and furthermore that the rate of expansion is accelerating.

Hence, the scientific consensus is that more space is being created around objects and this is how a greater distance is being created between objects. Much research over decades confirms this to be so, and both the redshift and Hubble's Constant come to mind.

It is a common misconception that redshift proves that galaxies are speeding away from us. They are not. Distant galaxies are not speeding through space. Space itself is expanding, putting greater distance between objects. It's a subtle difference, but it means that the galactic redshift is caused by cosmic expansion, not relative motion. Thus, cosmic expansion is the force that is creating inertia, and hence gravity.

As mentioned, when this additional space is created around objects it causes inertia to such objects from all directions. And remember that this is happening continually, all the time, right now. The inertia is caused by the fact that the creation of additional space is happening at a constant accelerated rate (not at a steady rate). So inertia is caused by the non-stop creation of new additional space around objects at an accelerated rate. This creates a pushing effect against objects from all directions:

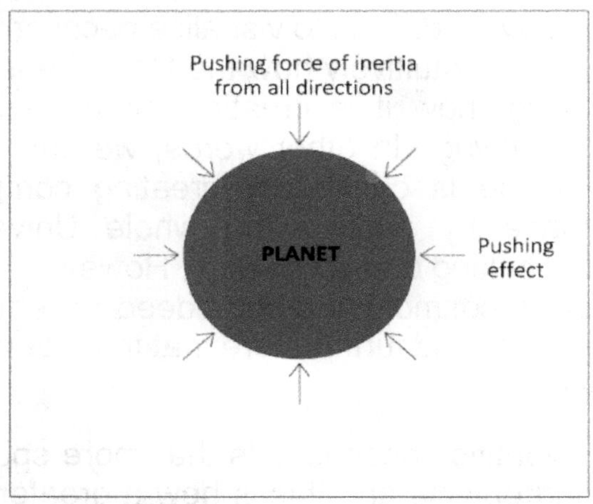

To clarify further, the inertia is caused by the *accelerated* movement (i.e. cosmic expansion) of objects as they move away from each other in all directions. To emphasise: the 'moving away' is in all directions, so inertia is applied for example to a whole planet in all directions (i.e. on all parts of the planet's surface).

Obviously, nothing can 'move away' in all directions. What happens is that as more space is created around, say a planet, it means there is more distance between the planet and other objects in all directions; so in this sense the planet has moved away from other objects in all directions.

But if objects are close enough to be affected by each other's gravity, the same pushing effect of inertial gravity will keep such objects together. Why so? Because as the AE force tries to separate objects in close proximity it will not be strong enough against the existing inertial gravity among the objects.

And the existing inertial gravity among such objects becomes one overall mantle of gravity (as if it was just one object). The AE force treats this one overall mantle of gravity as if it were one object. This keeps the parts of the 'one object' together, whether it be atoms in a body, different things on a planet, or stars in a galaxy.

This explains how galaxies are kept together and how greater distance is put between distant galaxies. More about this in the section 'What keeps galaxies together'.

So in the context of the fundamental forces of the Universe, there is only one type of gravity, and it is caused by inertia. This inertial gravity also works across gaps because gravity is a pushing force.

Note: The *same* AE force that causes inertial gravity in big objects also causes quantum gravity; this is covered later in the book.

You will know, for example, that if you leave an air gap between your human back and your car-seat, when the car accelerates you will feel inertia pushing you back into your seat, taking up the gap. This happens because the force of inertia caused by acceleration pushes the car from behind into your back.

So if two objects are close enough, such as your back and the car seat, inertia works over empty space. If you jump out of an aeroplane you will be close enough to Earth to be affected by Earth's inertia which pushes you towards Earth as you fall.

Thus, things that are close enough to Earth such as a parachutist, a hang glider, a satellite or the Moon will be affected by the inertia around all parts of the planet Earth. Why? Because gravity is a pushing force, it is pushing the parachutist, hang glider, satellite and Moon towards Earth (by virtue of Earth being the biggest mass in the vicinity).

When we consider that gravity is a pushing force, not a pulling force, many cosmological conundrums fall into place such as what keeps galaxies together and how were the first galaxies in the Universe formed.

For example, it is thought that galaxies were born when vast clouds of gas and dust collapsed under their own gravitational pull, allowing stars to form. Such reasoning is unlikely to be correct because in fact galaxies were formed relatively quickly after the big bang as evidenced by the latest research. In 2023 the James Webb telescope confirmed the existence of galaxy JD1 (first discovered in 2014), a galaxy thought to be over 13 billion years old.

Waiting for dust-cloud particles to bump into each other randomly, grow bigger, and develop gravitational pulling power would seem to take too long to account for the relative quick growth of galaxies after the big bang. Furthermore, a 'pulling' force of gravity has never been shown to exist as it implies that some kind of mysterious force (a separate agent) reaches out and pulls things inward.

But the pushing force of Universal Inertial gravity as revealed in this book is a different ball game. As the Universe expands it pushes objects together if they are close enough to be affected by each other's inertial gravity. What happens is that as the AE force tries to push cosmic dust particles apart, it makes them move more and bump into each other more often. In doing so, this greatly increases the coalescing rate of dust particles, leading relatively quickly to the formation of bigger objects and the formation of stars and galaxies.

It used to be thought that cosmic dust was formed in stars and is then blown off in a slow wind or a massive star explosion. But it is now thought that most cosmic dust comes from supernovae explosions. This helps to explain how stars and galaxies formed relatively soon after the big bang.

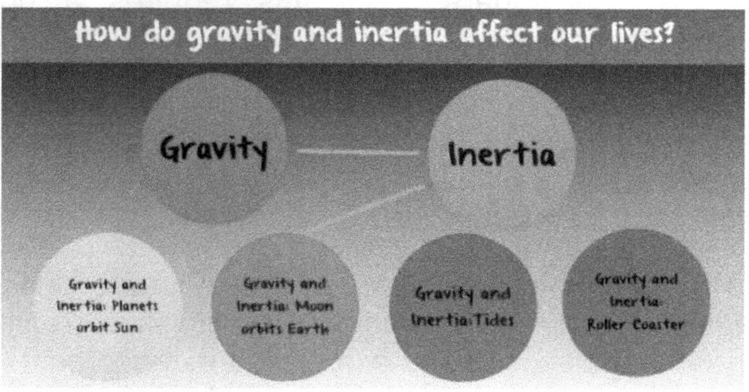

When it is argued that in empty space there is no means to produce a 'force' at a distance over other objects, this is so. But in the case of inertial gravity, it is a force that can indeed produce gravity between objects over empty space. It does this but not through some mysterious force that reaches out over empty space, it does this through movement. This movement is continually being created by the AE force, which is a pushing force.

Imagine a planet in space and that a small rock is floating nearby. As the AE force creates more space around the planet in all directions, the AE force exerts a pushing force on all parts of the planet's surface and a nearby rock will get caught up in the pushing force against all sides of the planet.

Thus the pushing force around the planet (with its greater mass) is much stronger than the pushing force around the mentioned small rock. So the small rock gets 'caught up' in the big pushing force around the planet and gets pushed towards the planet instead of moving further apart. This is how the AE force makes objects coalesce if they are close enough to be affected by each other's inertial gravity.

The bigger the mass the bigger the force of gravity. The force of inertial gravity is equal to the weight of mass multiplied by the acceleration squared. But this requires a point of clarification regarding the following conundrum.

If the Universe is expanding at an accelerated rate, and thus causing inertial gravity, why isn't the inertial gravity that we experience in our daily lives always becoming stronger and stronger? Here is the answer:

The consensus is that the force of gravity in the Universe is not changing, or at least changing too slowly to register on human time scales. Here is a recent announcement on the subject:

"According to a new study by the International Dark Energy Survey (DES) Collaboration, the nature of gravity has remained the same throughout the entire history of the Universe. These findings come shortly before two next-generation space telescopes (Nancy Grace Roman and Euclid) are sent to space to conduct even more precise measurements of gravity and its role in cosmic evolution". Source: DES findings presented in 2022 at the International Conference on Particle Physics and Cosmology (COSMO'22).

It is generally accepted that the expansion of the Universe is occurring at an accelerated rate. Distant galaxies are moving away from our milky way at a faster and faster rate. So this is the conundrum: It is known that acceleration causes inertial gravity, so if the rate of acceleration is increasing, this should make inertial gravity become stronger and stronger. But that is not happening. So why not?

The answer is that the rate of acceleration (of cosmological expansion) is constant. Imagine you are in a car and the speed is gradually increasing at a steady rate (every second the speed goes up by say 1 mph). In this scenario, the rate of increase in the car's acceleration is constant and consequently the force of inertial gravity on your body in the car seat will continue, but the strength of inertial gravity will stay unchanging.

So acceleration is what causes inertial gravity, and if the rate of acceleration is constant, the force of gravity remains constant (unchanging).

Regarding the acceleration of cosmological expansion, there may be some disagreement among scientists as to the actual rate of accelerated expansion. But all are agreed that whatever the rate of accelerated expansion may be, it is a constant rate of acceleration. The Universe is not chopping and changing its rate of accelerated expansion and that is why the fundamental force of gravity in the Universe has remained unchanged.

In the section 'The fundamental forces of the Universe' the four fundamental forces of the Universe are shown to be responsible for the creation of atoms and the Universe as we know it. But a fifth fundamental force is needed to bring together these four fundamental forces. The AE force does this by providing the common ground between the gravity we are familiar with and the gravity of subatomic particles:

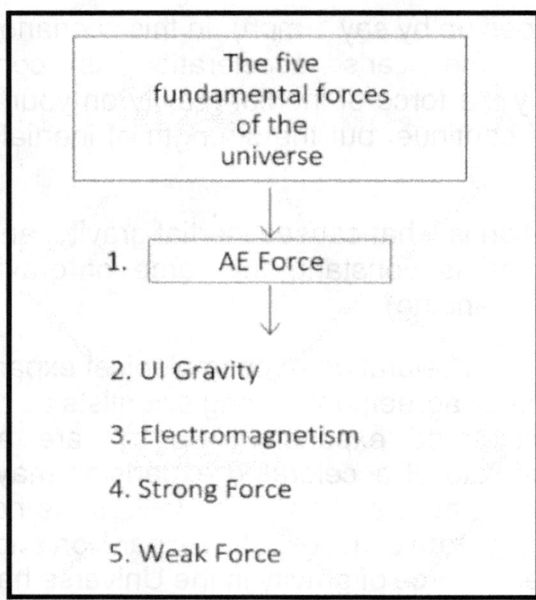

The five fundamental forces of the universe
1. AE Force
2. UI Gravity
3. Electromagnetism
4. Strong Force
5. Weak Force

In summary, this is how the AE force works:

1. The AE force (accelerating expansion force) exerts a force of inertia on all objects in the Universe, including subatomic particles.

2. The AE force of inertia is what we know as gravity, and is the cause of gravity in everything that exists, including subatomic particles.

3. The AE force creates a continual pushing apart force between and around all objects, including subatomic particles. The pushing apart force can be cancelled out by the pushing together effect of UI gravity if they are close enough to be affected by each other's gravity. This is what holds atoms together in the human body, or stars together in a galaxy.

4. The AE force is at the route of the way objects behave (and how their gravity is created), both on a micro level and on a macro level.

5. The AE force is continually pushing apart everything that exists thus imbuing everything with gravity. And most things that exist are simultaneously coalescing from the force of gravity created by the AE force.

Examples of how the AE force pushes objects apart:

* Planet Earth is pushed apart from distant galaxies in all directions as verified by the Hubble Constant and the redshift.

* The Milky Way galaxy is pushed apart from other distant galaxies in all directions.

* Our local cluster of galaxies (the 'M31 Cluster') is pushed apart from other clusters of galaxies in all directions.

Examples of how the AE force pushes objects together:

* The atoms in the human body are pushed (kept) together to form a whole body.

* The Sun and its local planets are pushed together to form the Solar System.

* The stars and planets in the Milky Way are pushed together to form a single galaxy.

* The Milky Way and a group of other galaxies in the vicinity are pushed together to form the M31 Cluster of galaxies.

The AE force is the master fundamental force of the Universe in that it provides the infrastructure for the other four fundamental forces to exist. But that is only half the story, the other half is the role played by gravity in all of this, which is what we will discuss next.

The take home message: The AE force refers to the accelerating expansion of the Universe. It provides inertial gravity to everything in existence and brings together the other four fundamental forces of nature to give a *Final Theory Of Everything*.

The underlying nature of gravity

In this section the underlying nature of gravity is examined and how it fits in with the AE force and the *Final Theory Of Everything.* Also, the nature of orbits and the concept of 'gravity mantles' is discussed.

If you ask the question: How far does Earth's gravity reach into space? Some physicists will incorrectly say that Earth's gravity waves extend to infinity into space, decreasing with the inverse square of distance from the object. This is baseless firstly because there is no evidence showing that gravity waves exist, secondly to say that something extends to infinity is simply uncorroborated conjecture, and thirdly gravity is not a force that 'reaches out into space' - gravity is always a pushing force caused by inertia as explained in this book.

It is claimed that Einstein's general theory of relativity correctly describes the fundamental nature of gravity to over 30 orders of magnitude, from submillimetre scales all the way up to cosmological distances. It is also claimed no other force of nature has been described with such precision and over such a variety of scales. Hence, with such a level of impeccable agreement with experiments and observations, general relativity would seem to provide the ultimate description of gravity.

In the aforementioned, the phrase '30 orders of magnitude' is meaningless in the context of measuring gravity accurately. Is this referring to a force of gravity? If so, what kind of force? Or is it merely a way of saying '30 times greater', and if so greater than what? And to say that 30 orders of magnitude describe a measurement of great precision, it begs the question: compared to what? And why don't advanced gravimetry instruments include general relativity to calculate gravity accurately?

Here is the Einstein Field Equation that is used to measure gravity with so-called 'great precision':

$$R_{\mu\nu} - \frac{1}{2}Rg_{\mu\nu} + \Lambda g_{\mu\nu} = \frac{8\pi G}{c^4}T_{\mu\nu}$$

By definition, this field equation must take into account the Einsteinian curvature of space. To do this a variety of 'tensor equations' are put into the equation, represented in the above image by R, g, and T. These three mathematical tensors are contrived sets of algebra to make the field equation work. In mathematics a tensor is simply some algebra whose purpose is to measure the geometrical shapes of objects such as a theoretical geometric curvature of space.

But the aforementioned Einstein field equation does not work on a cosmological scale and it certainly does not provide a measurement of gravity to an accuracy of '30 orders of magnitude'. Here are two main reasons why this field equation is spurious:

1. The equation attempts to set a gravity 'Constant' that can be applied to any part of the Universe. But we know that any measurement of gravity will vary from place to place depending on the prevailing factors of the moment, such as mass, energy, location, etc. A Constant 'G' that can be used for calculating gravity cannot exist - it would have to be something in the Universe that always gives the same mathematical result when its gravity is measured at any time (this does not exist).

2. The equation attempts to set a speed of light as a 'Constant' that can be applied to any part of the Universe when using the

formula for calculating gravity. But we know that the speed of light (the journey-time of a light ray from A to B) can vary depending on the prevailing factors of the moment such as the medium of the environment, the degree of light refraction, light frequency, etc. Hence, in the context of measuring gravity, a speed of light 'Constant' cannot be applied, but this is what the Einstein field equation attempts to do.

"There are the two considerable reasons, which undoubtedly show that Einstein's field equations are not correct from the point of view of physics, and this actually means that the search for their mathematical solutions is meaningless" (source: extract from: The special theory of relativity - the Biggest Blunder in Physics of the 20th Century, 2018, Gocho V. Sharlanov, Master of Science in Engineering, physics.bg).

Coming back to our Universal Inertial gravity as revealed in this book (let's call it UI gravity to distinguish it from Einsteinian spacetime gravity), the way UI gravity is calculated is well-established. The pushing effect of UI gravity on a free-falling object will yield about the same 9.8 newtons per kilo yardstick (9.8 metres per second for every second) in all parts of Earth's surface. We are not re-inventing the very well-established mathematics that describes inertial gravity.

To reiterate, inertia is a force created by any kind of accelerated movement. This explains how gravity, arising from inertia, is everywhere. And how gravity has a pushing together effect on things if they are close enough. If things are not close enough the same pushing effect pushes objects further apart as evidenced by the redshift, the doppler effect and the Hubble Constant.

You can think of it like this: The AE (accelerating expansion) force is continually creating more space, thus moving objects apart which in turn creates a pushing effect of inertia around objects. This pushing effect (i.e. gravity) is continual. If objects are close enough to each other the pushing effect continues pushing (keeping) them together.

If objects are not close enough to be affected by each other's gravity, such as the space between distant galaxies, the force of inertia separates the objects more and more because there are no nearby objects (clumps of mass) that can act as points of gravity to prevent the separation.

"The stars that make up galaxies are gravitationally bound to one another - so they do not come apart due to the force created by accelerating expansion. But if two galaxies are sufficiently separated - they will be carried apart by the force of expanding space" (source: Jen Jamison, Cosmologist & pioneer of inertial guidance systems, USA).

The third postulate of the FTOE states: Everything that exists in the Universe is continuously moving, either coalescing (moving together) or separating (moving apart). So in the context of gravity, all objects in the Universe either coalesce (move together) or separate (move apart). There is no other type of movement.

Everything without exception is either coalescing or separating (or doing both). The coalescing movement may be very gradual, over say millions of years, as in the case of so-called 'stable' orbits. If the movement is 'separating' it means that objects are moving apart from other objects in all directions because of the AE force.

For example, the Earth is gradually coalescing with the Sun, which in turn is coalescing with the Milky Way. And the Milky Way is gradually coalescing with our local cluster of galaxies (called 'M31'). But at the same time, the Earth (and the Milky Way) is separating (moving apart) from very distant galaxies as evidenced by the redshift.

Another example is light, which always moves in straight lines. As starlight travels through the void it is separating, i.e. moving away from its source of light as evidenced by the redshift. Or light can coalesce (move towards an object) if its direction of travel takes it towards an object.

Even a solitary rock or an atom located in deep space, far from everything will be coalescing or separating. If the rock (however small) is within a galaxy it will be coalescing with everything else within the galaxy. If the rock is nowhere near a galaxy or any other object, it will be separating from all objects in the Universe due to the AE force that is continually putting more space around the rock at an accelerated rate.

So everything that exists, however big or small, is either coalescing as a result of gravity, or is moving away if not close enough to be affected by nearby gravity.

Note: The concept of inertial gravity and the supporting mathematics is nothing new, and is embraced in the standard model of physics. But what is newly revealed in this book is the concept that the accelerating expansion of the Universe creates inertial gravity around objects as it continually creates more space around such objects.

For nearly a hundred years since the heyday of Einstein to the present, cosmologists have not realised that gravity is actually a consequence of cosmic expansion, caused by the AE force (the fifth fundamental force of nature). But now, in the 21st century cosmologists are beginning to understand that the underlying nature of gravity is cosmic expansion rather than Einsteinian spacetime curvature.

"Gravity is the inertial reaction of matter to spatial expansion" (source: Bruce Jimerson, From Here to Infinity, Kindle edition, 2017). In other words, gravity arises from inertia caused by

the AE force (caused by the accelerating expansion force of the Universe).

The Coalescing Force

You may be wondering how can the same AE force act to push objects apart and also to push objects together? As mentioned previously, if objects are close enough to each other to be affected by each other's gravity, then such gravity will trump the AE force of separation, thus keeping things together. But how exactly?

This is where kinetic energy comes into play. As any physics textbook will tell you, the kinetic energy of an object is the form of energy that it possesses due to its movement. It is defined as the work needed to accelerate a body of a given mass. Having gained this energy during its acceleration, the body maintains this kinetic energy unless its speed changes. Put simply, kinetic energy is the movement-energy of an object.

As the Universe expands at an accelerated rate, the expansion itself keeps everything moving, whether it be moving apart or moving together. This AE force of universal movement creates kinetic energy in all objects because movement itself is what creates kinetic energy. Clearly, kinetic energy would be insignificant or undetectable at small/slow movements.

Kinetic energy is directly proportional to the mass of the object and to the square of its velocity. So the greater the kinetic energy, the greater the mass and vice-versa.

When it comes to UI gravity, we have the following four points to consider:

1. Accelerated movement is created by the AE force and this gives objects kinetic energy. So as more space is created by

cosmic expansion, this increase in cosmic space makes objects move in all directions (like the raisins in the dough of bread in the oven). This moving apart gives such objects kinetic energy by virtue of the actual movement.

If objects are close enough to be affected by each other's inertial gravity, the AE force will nevertheless exert a force of separation on such objects. This in turn makes such objects move and acquire greater kinetic energy, but because they are near each other the kinetic energy serves to increase their force of inertia (gravity) on each other, thus ensuring that they continue to stay together. Colloquially, the objects get caught up in each other's inertial gravity and are pushed or kept together.

2. The kinetic energy arising from the movement created by the AE force increases the mass of such moving objects.

3. The greater the mass, the greater the inertia. And vice-versa, the greater the inertia, the greater the mass (they go hand in hand).

4. UI gravity (Universal Inertial gravity) arises from steps 1 to 3: accelerated movement of object → kinetic energy of object → greater mass of object → greater inertia of object → greater UI gravity.

So when it is said that the AE force pushes objects together when they are close enough to be affected by each other's gravity, this coalescing effect can be rephrased as follows:

The coalescing effect is caused by the AE force which acts to make objects move. This movement maintains or increases kinetic energy and in so doing it maintains or increases UI gravity thus keeping objects together if they are close enough to be affected by each other's gravity. This is how the AE force pushes objects together.

When objects are close enough to be affected by each other's gravity, it means that such objects will be pushed or maintained together by virtue of being within each other's gravity mantles. The coalescing force works through kinetic energy and gravity mantles.

Gravity Mantle

Everything that exists has a gravity mantle. For example the planet Earth has a gravity mantle that extends into space beyond the moon. The gravity mantle surrounds its source object, and the bigger the mass of the object the bigger its gravity mantle.

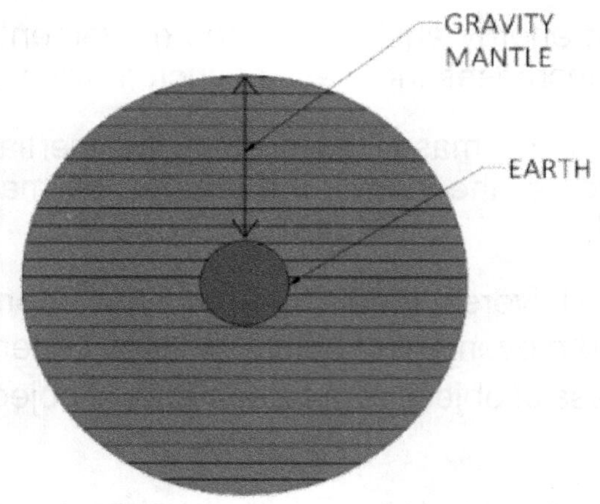

Think of the gravity mantle as the area of space in which the gravity of an object is manifested in the sense of being significant and measurable. Since gravity is caused by inertia (emanating from the AE force), the gravity mantle represents the space around an object that manifests inertial gravity.

So, for example, if you are above Earth, but within the gravity mantle, you will experience Earth's gravity. The further away you go from Earth the weaker the force of gravity. Beyond the gravity mantle of an object, the force of gravity becomes too weak to detect or too negligible to measure.

There is no precise mathematical cut off point for calculating the gravity mantle perimeter or its radius because the force of gravity fades away very, very gradually the further you go away from an object. Eventually gravity becomes too weak or non-existent to measure or be significant, but to give you a very rough idea, the radius of a gravity mantle can be anything between 20 times to millions of times larger than the radius of the object producing the gravity mantle.

The mass of a star is so much bigger than the mass of a planet and therefore the gravity mantle of a star is going to be millions of times bigger than the gravity mantle of a planet.

As we have said, the gravity mantle is created by the new space that is always being created around objects as a result of cosmological expansion. This AE force pushes objects apart if they are not close enough to be affected by each other’s gravity.

But if objects are close enough to be affected by each other's gravity they will coalesce. This coalescing occurs when gravity mantles overlap. The overlapping of gravity mantles means that an object enters the area of gravity of another object (or objects) and thus such objects fall prey to each other's gravity.

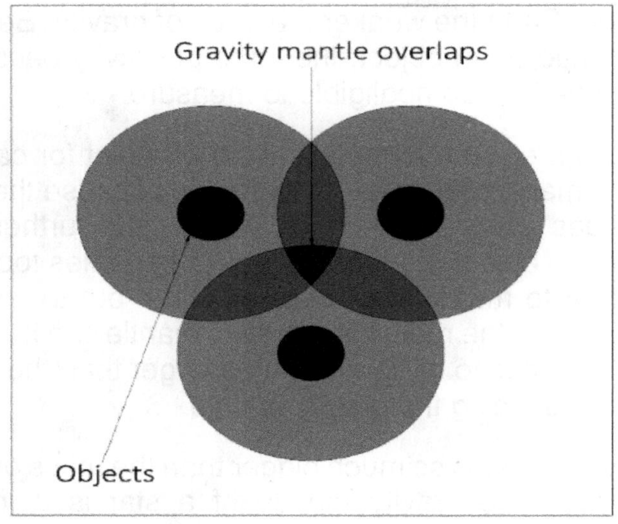

We should understand that the force of inertial gravity does indeed work across empty space. This 'force' is not an external agent, it is simply movement. Let us remember the example of an air gap between your human body's back and your car seat. If you move forward in your seat when the car is stationary (so as to leave empty space behind your back), then as the car accelerates you will feel inertia pushing you back against the car seat. So the movement of the car has created a force of inertia in the air gap between your back and the car seat.

The point here is that the force of inertia works across empty space. For those physicists who say that inertia is not a force but a resistance to a change in motion, it should be stated that this is so. The inertia that you feel in the car seat is the resistance of your body to a change in motion as the car accelerates. We loosely use the word 'force' as shorthand for saying 'a resistance to a change in motion'.

The force of gravity on Earth comes from Earth's resistance to a change in motion caused by cosmic expansion as it tries to move Earth away from other objects. Earth is able to resist this change in motion because its gravity mantle is completely within the Sun's gravity mantle. Gravity mantles can overlap in part or overlap completely depending on mass, location, etc.

When you fall off a chair and hit the floor it happens because your body's own gravity mantle is within Earth's gravity mantle. If you are falling to Earth in a parachute the same applies.

What about rocket man? If you're in a rocket and halfway to Mars, the gravity mantle of the rocket will be within the gravity mantle of the sun and may also overlap with the nearby gravity mantles of planets in the solar system, thus affecting space navigation to Mars. If the rocket accelerates, its speed will create temporary 'artificial' gravity on, say, the floor of the rocket. This artificial gravity will have its own small gravity mantle within the rocket and it will overlap with the gravity mantle of anybody standing on the floor of the rocket, thus creating an area of inertial gravity within the rocket.

Note: When gravity mantles overlap, a so-called 'gravity barycentre' is formed. *"A barycentre is the centre of mass of two or more bodies that orbit one another* [or that come close to one another] *and is the point about which the bodies will orbit* [or coalesce]. *A barycentre is a dynamical point, not a physical object"* (source: Barycentre, Wikipedia.org). The barycentre moves around depending on the movement of the objects, and it can be located anywhere within the mantle overlaps or within (inside) one of the objects itself.

So the force of inertial gravity is caused by a resistance to a change in motion, and when gravity mantles overlap (partially or completely), objects coalesce instead of separating. Every object in a galaxy has its own gravity mantle which overlaps with other nearby gravity mantles, and they in turn overlap with others. Thus all the objects in a galaxy form a galaxy-wide 'network' of connected gravity mantles that keep everything together.

Subatomic particles also have gravity mantles, but they are too weak to account for quantum gravity, so other factors come into play as explained in the section 'Quantum Gravity' that follows shortly. **Note:** More about gravity mantles in the section 'What keeps galaxies together'.

Orbits

Orbits in the cosmos and orbits in subatomic particles behave in a similar way. For example, in the case of an atom, the nucleus is made of protons. And a proton is about 1,800 times the mass of an electron. So the much smaller electrons orbit the protons, and not the other way round. The greater gravity of a proton makes the electron be the orbiting object.

In quantum mechanics scientists talk about 'allowed' orbits. That is, subatomic particles cannot orbit in any arbitrary way. They have to follow 'allowed' orbital paths that suit the energy profile of the orbiting particle. If the orbiting particle were to change to a different orbit, the composition of the atom would change by virtue of having different orbits. This in turn can change the nature of an atom into a different kind of atom. Nevertheless, although subatomic particles follow allowed orbits, the orbits occur as a result of quantum gravity, the nature of which is covered later in the section 'Quantum Gravity'.

In space the orbit of an object around its 'parent' is a balance between the force of gravity and the object's desire to move in a straight line arising from its momentum. An orbit occurs when the object possesses just enough speed (momentum) to push away slightly from its parent (but not escape it). This causes the orbital speed to reduce, so eventually the object will be moving slow enough to be pushed back in. Hence, the object's distance from its parent oscillates, resulting in an elliptical orbit.

All orbits in the Universe, whether it be planets around the sun or electrons around an atom, have the following in common:

1. All orbits are elliptical, never a perfect circle.

2. All orbits undergo a precession (a change in orbit orientation). The path of the next orbit will not be exactly the same as the last orbit.

3. All orbits are coalescing, i.e. they are decaying and becoming weaker. With enough time orbits will collapse, making the orbiting object coalesce (fall) to the object being orbited.

4. All orbits are caused by the orbiting object falling toward a more massive object, but in the case of orbits, its speed and mass is such as to make the object follow a curved tangential path.

A major reason for orbits to remain relatively stable (but never completely stable) relates to kinetic energy. Planetary orbits are an excellent example of the principle of energy conservation, because over time, a mass in an elliptical orbit converts energy from potential to kinetic and the reverse, while maintaining a near constant level of energy (the sum of potential and kinetic energy).

When an object, say a moon, goes past a planet at high speed, the proximity to the planet pushes the moon inward with enough force of gravity to change its direction into a tangential path. Put another way, the trajectory of the moon takes it past the planet and into the planet's gravity mantle, but the momentum of speed makes the moon continue moving out of the planet's mantle, but on a tangential path as a result of dipping into the planet's gravity mantle.

This slows the moon down such that it again temporarily falls prey to the planet's force of closeness, and the process is repeated.

Think of it like this: the moon's speed makes it want to move off in a straight line, however the earth's gravity moves it slightly out of line, making the moon follow an elliptical orbit.

Each elliptical orbit becomes a little slower (weaker) than the previous orbit because of gravity, and eventually (all things being equal) the orbiting object will fall into the object being orbited. So all orbits, including subatomic orbits, are coalescing. But on human time scales orbits can appear to be stable because some orbits can take millions of years to decay.

"Every orbit — even gravitational orbits in General Relativity — will very, very slowly decay over time. It might take an exceptionally long time, some 10^150 years, but eventually, the Earth (and all the planets, after enough time) will have their orbits decay, and will spiral into the central mass of our Solar System". Source: Ethan Siegel, How Our Solar System Will End In The Far Future, Forbes Science, Jan. 2017.

Our moon, however, is the exception that proves the rule. The moon is continually being pushed into a higher orbit by ocean tidal forces on Earth, and is moving away from the earth at nearly 4 cm per year. But even this is temporary. When the oceans dry up on Earth (in about a billion years), the moon will start to coalesce with Earth.

The principles that control how moons and planets orbit also apply to entire galaxies and groups of galaxies. The AE force continually creates space and hence UI gravity (Universal Inertial gravity) around everything within a galaxy. But within a galaxy, things coalesce because the smallest objects will be close enough to bigger objects, and they in turn will be close enough to even bigger objects, and so on up the food chain to the biggest objects in the galaxy.

By being 'close enough' it means the force of gravity trumps the force of separation on a local galaxy-wide level. In other words, when gravity mantles overlap they are close enough to be affected by each other's gravity. So all objects within the galaxy continue coalescing and continue having gravity. But the galaxy as a whole will be falling prey to the force of separation on a large cosmological scale.

Why does everything in the galaxy orbit together in a kind of spiral movement? Cosmologists explain that this is because the immense combined mass of the galaxy, most of it near the centre, creates immense gravity that pulls all the stars (and their accompanying planets) into elliptical orbits.

This explanation is unlikely to be correct because there is no 'immense gravity' that keeps everything in the galaxy together. Such 'immense gravity' has never been detected. A much more likely explanation is that, as explained, the gravity mantles of objects in a galaxy are inter-connected.

Think of it like this: a hang-glider is pushed to Earth because Earth's force of gravity trumps the force of separation. Coalescing wins out over separation because the hang glider was close enough to Earth for this to happen. It's the same among the stars and planets within a galaxy - they are close enough to coalesce.

And it's the same for a cluster of galaxies. On a cosmological scale, every galaxy will have its own overall gravity mantle.

When the gravity mantles of galaxies overlap it means they are close enough to coalesce instead of falling prey to the AE force of separation.

How far must you travel to escape Earth's gravity mantle? An altitude of 100 kilometres is considered as the border between the atmosphere and outer space; once the spaceship is beyond this altitude it is in a state of freefall (weightlessness) but not zero gravity. This altitude is known as the Karman line. Beyond that, at some point the force of Earth's gravity (way beyond the moon) becomes negligible or too small to measure.

It is claimed by some that gravity is never zero anywhere in the Universe; that the force of gravity decreases by $1/r^2$ so it is felt by objects at all corners of the Universe. **Note:** The coefficient 'r' in the aforementioned sentence is a measurement of statistical accuracy.

In fact it is likely that gravity is not everywhere in the Universe. There are parts of the Universe where there is probably no gravity, i.e. no force of gravity that can be detected. All objects in the Universe have gravity, but for example, there are vast spaces between galaxies where there is no gravity (or no detectable gravity). In other words, gravity does not extend beyond galaxy groups or exist between distant objects. If gravity existed everywhere there would be no redshift showing that distant galaxies are moving away from each other.

If you look at Newton's Law of Universal Gravitation, you see that the force of gravity on one mass due to another mass depends on their separation 'r' according to the dependence 1/r2. As you get farther away from a gravitational body such as the sun or the earth (i.e. as your distance r increases), its gravitational effect on you weakens but never goes completely away; at least according to Newton's law of gravity.

Although Newton's Law of Gravity is quite accurate, it is not entirely so because it is thought that *"the effect of gravity is not infinite"* (source: Dr Christopher Baird, Theoretical Physicist, Does the influence of gravity extend out forever? wtamu.edu, June 2015).

The standard model considers general relativity to be the most correct model providing you add dark matter to the theory. But as mentioned, dark matter has never been detected and remains unproven. It is a modern-day equivalent of the mystical 'ether' as first proposed around 1704 - a substance that is said to fill empty space everywhere.

General Relativity states that gravity occurs when the mass of an object triggers spacetime to make space curve around such a mass causing objects that are in that curved space to go towards the bigger mass.

In comparing Newton's Law of Gravity and Einstein's Theory of Gravity, neither is correct, although Newtonian physics rules the day in modern science in all types of space navigation (more about this in the section 'Space Navigation'). If Einstein's Theory of Gravity were correct it would have to mean that all parts of the Universe exist as curved space because Einsteinian gravity requires curved space for gravity to exist - clearly this is absurd.

The take home message: The underlying nature of gravity is revealed as inertial gravity that is caused by the AE force. The movement of things caused by cosmic expansion (the AE force) gives kinetic energy to everything. So movement and kinetic energy work together to give gravity to everything in the Universe.

Quantum Gravity

This section explains the next part of the puzzle that makes up the *Final Theory Of Everything*: quantum gravity. Also, the surprising fundamental nature of black holes is revealed for the first time.

When it comes to quantum gravity, many books have been written on the nature of subatomic particles, how atoms and molecules behave and other subatomic phenomena including quantum gravity. We won't be repeating such well-studied and researched information here. Rather, the focus is on the nature of quantum gravity and how this fits in with the *Final Theory of Everything* as postulated in this book.

Quantum gravity has been a subject of research for many years as scientists try to reconcile micro and macro gravity; it is the holy grail of contemporary theoretical physics. Researchers have come up with ideas for how they might find clues to its existence by looking at black holes, the early Universe and other cosmological phenomena, but no one has yet turned up any hints of quantum gravity in nature.

"Gravity is the weakest, and yet the most pervasive, of the four basic forces is gravity. It acts on all forms of mass and energy and thus acts on all subatomic particles, including the gauge bosons that carry the forces" (source: Britannica.com, Gravity).

"Reconciliation of general relativity with the laws of quantum physics remains a problem as there is a lack of a self-consistent theory of quantum gravity. It is not yet known how gravity can be unified with the three non-gravitational forces: strong, weak and electromagnetic" (source: Wikipedia.org, General Relativity).

Today, many scientists suspect that the common denominator

between macro and micro gravity will be found in quantum field theory (a theory that combines classical field theory, special relativity, and quantum mechanics). But to date no one has succeeded in explaining and showing how gravity works at a subatomic-scale even though it is known that subatomic particles are subject to gravity.

String Theory

The American physicist Richard Feynman was a pioneer in the search for a quantum field theory that could explain quantum gravity. His famous 'Feynman diagrams' come to mind which revolutionised theoretical physics and won him the Nobel Prize. He is well-known for saying: *"I think I can safely say that nobody understands quantum mechanics."*

Shortly before his death Feynman admitted his scepticism of string theory by saying: *"I don't like that (with string theory) they are not calculating anything. I don't like that for anything that disagrees with an experiment, they cook up an explanation"* (source: Lefteris Kaliambos, False Feynman Diagrams, Fundamental physics concepts).

"In his later life, Feynman was critical of string theory, believing that it was a research dead-end" (source: The String Theory Debate, Massachusetts Institute of Technology).

String theory is the idea in theoretical physics that reality is made up of infinitesimal vibrating strings, smaller than atoms, and that these tiny vibrating strings can explain the nature of gravity for both big things and small subatomic particles. String theory postulates that any theory of everything would need to bring together Einsteinian gravity and quantum gravity into a single theory. Among other things, string theory requires the existence of up to ten dimensions (7 more dimensions than the 3 we are familiar with). Also, to make string theory work it requires the incorporation of the general theory of gravity. In short, string theory remains baseless and unproven.

For the sake of completeness 'M theory' should also be mentioned. Postulated in the late 1990's by Edward Witton, M theory attempts to bring together the various different string theories into one umbrella theory comprised of 11 dimensions. M theory lacks a complete mathematical formulation and is still being developed.

In his life Feynman was always sceptical about Einsteinian relativity and he spent much of his life trying to find ways around it. *"The existing maths as used in Special Relativity is fatally flawed. The only person that appears to have tried to correct this flaw is Richard Feynman"* (source: Roger J. Anderton, General Science Journal)".

Albert Einstein once attended a lecture given by Feynman. Afterwards Einstein commented that Feynman's ideas were inconsistent with the principles of general relativity. But that, he said, wasn't necessarily so bad *"after all, general relativity is not so well-established as electrodynamics."*

As mentioned, it is postulated in this book that there is only one type of gravity in the macro Universe: UI gravity. And one type of gravity in the micro Universe: electromagnetic gravity. Both types of gravity are born from the same AE force. It would not be correct to suggest that there are two fundamentally different incompatible types of gravity, one for the macro and one for the micro, with no common parentage.

We have also said that everything that exists is either coalescing (coming together) or separating (moving apart), or doing both depending on the perspective. We know for example, that stars in the Milky Way are gradually coalescing, and we know that simultaneously the Milky Way as a whole is separating from other distant galaxies in all directions.

Why hasn't quantum gravity been reconciled with the standard model of Einsteinian gravity? The answer is that it cannot be reconciled because Einsteinian gravity (i.e. General Relativity) is baseless, a falsehood based on space curvature, and attempts to reconcile the two is a forlorn hope.

If you try to investigate exactly why there is no reconciliation a typical answer goes something like this: The smooth, continuous Universe of general relativity describes conflicts with the discrete, chunky Universe of quantum physics. When you bring their equations together you get nonsense.

Wikipedia describes quantum gravity like this (abridged extract):

"Quantum gravity describes the principles of quantum mechanics. Three of the four fundamental forces of physics are described within the framework of quantum mechanics and quantum field theory. The current understanding of the fourth force, gravity, is based on Albert Einstein's general theory of relativity, which is formulated within an entirely different framework, such as spacetime curvature. There is a need for a theory that goes beyond general relativity into the quantum realm, so as to allow, for example, an understanding of physics at the centre of a black hole".

In comparing what we know about quantum gravity with UI gravity, a good starting point is to look at the three postulates of the FTOE (*Final Theory Of Everything*) which are:

Postulate one: *Everything that exists has mass and is subject to gravity. Stars and galaxies are subject to inertial gravity, and atoms/subatomic particles are subject to electromagnetic gravity; both types of gravity are born from the same AE force.*

Postulate two: *The force of inertial gravity is caused by the constant accelerating expansion of the Universe (whatever the rate of acceleration may be), making objects come together, go into orbit or move away from each other depending on their proximity.*

Postulate three: *Nothing that exists can be motionless or 'at rest'. Everything that exists in the Universe is continuously moving, either coalescing (moving together) or separating (moving apart).*

The AE force complies with these three postulates. The accelerating expansion makes objects move apart by creating more space between them, and in so doing creates a pushing force of inertial gravity for everything that exists. This AE force keeps everything moving in the Universe.

This same AE force also causes inertia around subatomic particles, but it is too weak to explain the quantum gravity that we see manifested in such particles. If quantum gravity is not caused by inertia, then how is quantum gravity created? The answer is that the AE force causes quantum gravity by making subatomic particles move. It makes them move in the same way that the AE force makes everything else in the Universe move.

The human body is a walking collection of atoms. But in fact the human body is 99.9999999 empty space. If we lost all the dead space inside our atoms, we would each be able to fit into a particle of dust, and the entire human species would fit into the volume of a sugar cube.

The point is, every atom in our body is surrounded by empty space, and as the Universe expands, the AE force (accelerating expansion force) tries to add more empty space around atoms. In so doing, this exerts a force of separation, thus exciting subatomic particles into greater movement and greater kinetic energy. It is this force of movement that creates electromagnetism, and this in turn endows subatomic particles with quantum gravity.

So the AE force gives gravity to atoms without actually forcing them apart. The proximity of atoms to other atoms ensures their combined gravity keeps them together, i.e. keeps them coalescing, just as the stars in the galaxy are coalescing.

The idea of empty atoms huddling together, creating our bodies (and things such as buildings and trees) might seem a little far-fetched. To us those groups of atoms seem entirely solid, but on an atomic scale the human body is ethereal and gossamer (excuse the poetic indulgence).

If living things are so ethereal, full of empty space, how did they come into being you might well ask? The first organic molecules were very simple carbon-based molecules made of few atoms. These molecules then combined with other simple molecules to form more complex molecules. Over many years and probably trillions and trillions of chemical reactions, more complex and stable molecules were formed. The next step went like this:

The emergence of life from non-life certainly occurred. The forces of electromagnetism and gravity in the presence of complex molecules, is all that's required. As far as science can tell, what makes living things come into existence is simply the presence of electricity: the flow of electrons.

"With just the four fundamental forces in the Universe (gravity, electromagnetism, and the strong-and-weak nuclear forces) we can form atomic nuclei, atoms, molecules, life, complex-and-differentiated life, where consciousness emerges and some of those conscious beings can study the Universe itself" (source: Dr Marcelo Gleiser, How do fundamental particles create consciousness? bigthink.com, Sept 2022).

All subatomic particles are moving constantly, never stopping. A subatomic particle is by nature a moving particle; if it were not moving it would not be a subatomic particle. Like stars and galaxies, subatomic particles are gradually coalescing (coming together). This coalescing causes subatomic particles to stay together and eventually to decay or change to other types of subatomic particles.

And of course, subatomic particles located in our bodies and in our galaxy are also continuously separating from subatomic particles located in distant galaxies (in all directions).

For the technically minded this is how atoms can and do coalesce:

Given the fundamental forces that control the stability of atoms, there exists a ratio of protons and neutrons in the nuclei of all atoms that allows those atoms to be 'stable'. As the ratio in any given atom departs from this 'ideal', the nucleus becomes increasingly unstable and it spontaneously decays radioactively. As the atomic weight increases above 110, this instability becomes greater until the nuclei can only exist for small fractions of a second before decaying.

In brief, atoms can 'randomly' shoot off an electron, spontaneously break down into smaller particles and coalesce with nearby subatomic particles to form different atoms or particles. This subject is well-studied and it is known that when atoms are excited (made to move more quickly) through heat, electricity or light, they are more likely to collide with other atoms, lose electrons, or break down in some manner.

Also, let us remember from the section 'The coalescing effect' that all subatomic particles have kinetic energy in varying degrees, thus helping to keep subatomic particles together.

It is often said that Einsteinian gravity cannot be reconciled with quantum gravity because the mathematics don't agree. In the media this subject is referred to as a *'battle between relativity and quantum mechanics because both are fundamentally different theories that have different formulations'*.

It is not just a matter of scientific terminology; it is a clash of genuinely incompatible descriptions of reality. The conflict between the two halves of physics has been brewing for more than a century. Basically you can think of the division between relativity and quantum systems as 'smooth' versus 'chunky'.

This 'smooth' nature of general relativity contrasts sharply with the 'probabilistic' nature of subatomic particles. However, in reality the outcomes (positions/locations) of subatomic particles are not 'probabilistic' at all. But because of the very high speed and very small size of subatomic particles, we humans are not yet capable of predicting where a particular particle is located, where it will move to or what it will do. We cannot 'determine' the behaviour of a subatomic particle so we say it is not a 'deterministic' science. We simply do not have the technology or capability of doing so and perhaps we never will. But that doesn't mean there is no cause and effect, as there is in the macro world.

So scientists quite rightly adopt a probabilistic strategy instead of a deterministic strategy to study and explain subatomic particles; it's the nearest we can get to understanding the subatomic world.

Whole books have been written on this subject. Rather than devoting time to showing how the current theories of relativity and quantum gravity are both wrong, we will limit ourselves to showing how the AE force, as described in this book, imbues all subatomic particles with quantum gravity.

Subatomic Entanglement

There is a belief among some scientists that under certain conditions subatomic (quantum) particles can communicate between themselves instantaneously and over big distances. A repeatable experiment showing this to be so has not been successful to date.

I will let somebody else describe the issue:

Quote

You have two lights. One can flash red, the other can flash blue. You put one in a box, and the other in another box. You send ONE of the boxes to Pluto. When you open your box on Earth and trigger the light and see that the light is flashing red, you KNOW that the box on Pluto has the blue light. This implies 'instantaneously' and 'faster than the speed of light.

Here's the issue: no information is actually passed between the lights. The lights are 'collapsed' before the 'measurement' of opening the box. We were messing up in the initial state preparation, by 'knowing' that one light will be red, and the other blue, and collapsing them before they are sent. We interfered in the original creation of the 'entanglement', saying that one light should be red, and the other should be blue. Since we interfered in the initial state of the lights, we get misled into believing there is faster than the speed of light information transfer, when actually we knew the entanglement all along (source: ANASTASIA MARCHENKOVA, quantum telecommunications researcher and coder at the University of Maryland, USA).

Unquote

Simply put, so-called 'entanglement' of subatomic particles does not exist. The truth is probably more prosaic. The act of setting up an experiment to determine the state of an entangled particle is complicated, with all sorts of variables that can affect what you eventually see or record. Things like variations in gravity, ambient temperature, the state of instrumentation used, time of day, etc can all affect what you see when you look at a particular instance of particle entanglement.

Setting up two identical experiments on different parts of the planet is nigh impossible. And even if it were possible, it is very doubtful that some kind of mysterious faster than light communication between separated particles would emerge. Therefore one should be very sceptical when it is claimed that if particle A is changed in New York, this change affects particle B in London instantaneously.

Black Holes

In giving consideration to the conflict between relativity and quantum gravity, the subject of black holes is often mentioned. In what follows, the nature of black holes is discussed in light of the *Final Theory Of Everything* as postulated in this book.

Stars are born, they live out their lives and they die. They do this on time scales that can stretch to many millions of years. Stars are formed from an accumulation of gas and dust that coalesces due to gravity. The process of star formation takes around a million years from the time the initial gas cloud starts to collapse until the star is created and shines like the Sun.

The AE force previously discussed gives gravity to everything that exists, including the gas and dust that gives birth to stars. All stars eventually run out of their hydrogen gas fuel and die. The way a star dies depends on how much matter (i.e. mass) it contains. As the hydrogen runs out, a star with a similar mass to our sun will expand and become a red giant.

It is thought that when stars die some of them end up as a white dwarf, and then later as a black dwarf. However, some stars don't die 'quietly'; before they die they turn into black holes. It is thought that most black holes are formed from the remnants of large stars that die in supernova explosions. Each little bit of remnant from the explosion will be super-hot and dense, and hence have a large heat mass.

As we have said, the AE force gives inertia (i.e. gravity) to planets and stars, and everything else that exists, and this includes black holes. The greater the mass of an object, the greater the strength of gravity.

This high degree of mass of each remnant (from a supernova explosion) means a high degree of gravity. This in turn makes the remnants coalesce and come together spinning in what astronomers call a 'gravitational collapse into itself'. Because of the high amount of mass, and hence gravity, the collapse or joining of the supernova remnants continues until it turns into what we know as a black hole.

General relativity predicts that a sufficiently compact mass can deform spacetime to form a black hole. In other words, black holes are created or 'facilitated' by relativity. The reality is more prosaic: black holes are simply very dense agglomerations of matter. But simple it isn't. General relativity's equations fail catastrophically at a black hole's centre, known as a 'singularity', where the supposed warping of space-time simply goes off the scale. Put simply, general relativity is incompatible with black holes.

However, the nature of a black hole fits in nicely with the *Final Theory Of Everything* as postulated in this book. When a supernova explosion occurs it leaves behind a bunch of very dense remnants. They are dense because a supernova explodes when it becomes too dense. These dense remnants will readily come together because higher density equates to higher mass, and higher mass equates to higher gravity.

Supernova remnants are extremely dense on a par with our Sun being compressed into the size of an object smaller than 10 km in diameter.

"The ambient density and energy of supernova remnants are estimated from the intensity ratio of sulphur lines I(6717)/I(6731). It is found that, on average, the ambient density around galactic supernova remnants is 4 per cu cm. The total energy appears to be the same for all supernova remnants (to within a factor of about 5). A mean value of 4 by 10 to the 51st power erg is found. On the density and energy of supernova remnants" (source: Canto, J, Astronomy and Astrophysics, vol. 61, no. 5, Dec. 1977, p. 641-645).

As the supernova remnants coalesce they will begin to spin together, faster and faster, and this in turn produces immense kinetic energy. The combination of highly dense matter spinning faster and faster and gaining immense kinetic energy produces stronger and stronger gravity.

This immense gravity of a black-hole spinning very fast does not exert any kind of gravity-pulling power at all. This is a big misconception. But you, dear reader, will know from reading this book that black holes will have an immense gravity mantle by virtue of their immense gravity. Hence any object within the black hole's gravity mantle will be pushed towards the black hole, just like a parachutist is pushed towards Earth.

So as the supernova remnants coalesce into a black hole, and as the black hole gains super gravity, nearby objects will be pushed into the black hole, making its gravity stronger still. As this happens the black hole will grow bigger with time. In science the word for this growth is 'accretion' which describes how objects coalesce to form stars and planets (and black holes).

Due to the angular momentum of the spinning, a kind of vicious circle ensues inside a black hole in which the spinning and the gravity feed on each other until the speed of spinning nearly approaches the speed of light. When this happens it is theorised that as the black hole has reached its limit of spinning speed (and strength of gravity) it gradually throws out energy until eventually the black hole just 'evaporates' and 'peters out'. Here is a more technical description of this process:

"If angular momentum is to be conserved in a black hole, everything inside the black hole has to spin up their rotational speeds until they almost reach the speed of light. At that point, gravitational waves will kick in, and some of that energy (and angular momentum) gets radiated away. If not for that process, black holes might not be black after all. In this Universe, black holes have no choice but to rotate at extraordinary speeds". Source: Ethan Siegel, astrophysicist, This Is Why Black Holes Must Spin At Almost The Speed Of Light, August 2019, Forbes Science.

In the above quote Ethan Siegel states that 'gravitational waves' will kick in at a certain point. It is postulated that gravitational waves do not exist, but the very fast spinning of a black hole creates very strong inertial gravity - like spinning a bucket of water around your head, tied to a rope. The water stays in the bucket because of inertial gravity. It's the same inside a spinning black hole.

The very fast spinning creates very strong inertial gravity that gobbles up just about anything coming close to a black hole. As an analogy, think of twirling a bucket of water around your head tied to a rope. You twirl the rope so fast that when a fly accidentally goes into the twirling bucket it cannot escape from the inertial gravity.

What about light being sucked into black holes thus making black holes be black? This is a misconception. Light cannot be sucked into black holes just as nothing else can be sucked in. Every ray of light in the Universe expands as it moves forward, akin to a growing cone, as shown in the following image from left to right:

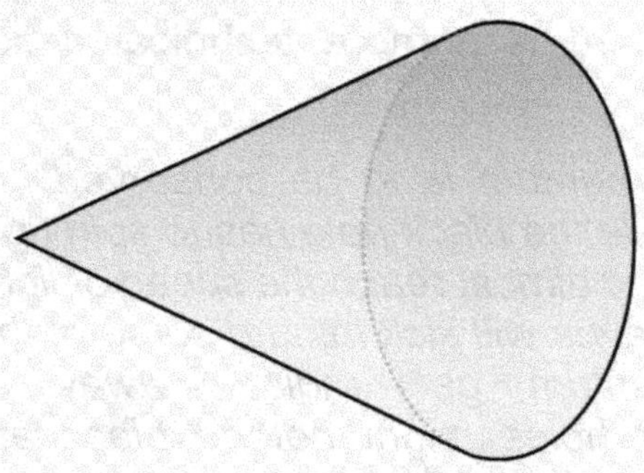

As the light moves forward, the photons that make up light become more and more dispersed, but always moving in straight lines. So when light passes near a black hole some of those dispersing photons may inadvertently go in straight lines to a black hole. Since light cannot bend or fall prey to gravity, the photons that happen to go to a black hole will continue straight on into the black hole and be destroyed. This is why we just see blackness when looking at a black hole.

When it is said that a black hole is so dense that even light cannot escape, this is not so. Light going into a black hole is destroyed, hence it does not 'escape' or come out again. A black hole is black because it emits very little energy compared to any other known objects in the Universe. Planets and asteroids for example emit ample energy in the form of refracted ('reflected') light from nearby stars.

With black holes it's different simply because they are so dense. So how does any amount of energy manage to escape from a black hole? Here is an explanation.

In the following image we see an ergosphere. This ergosphere is the 'entrance hall' or 'lobby' to a black hole before arriving at the event horizon:

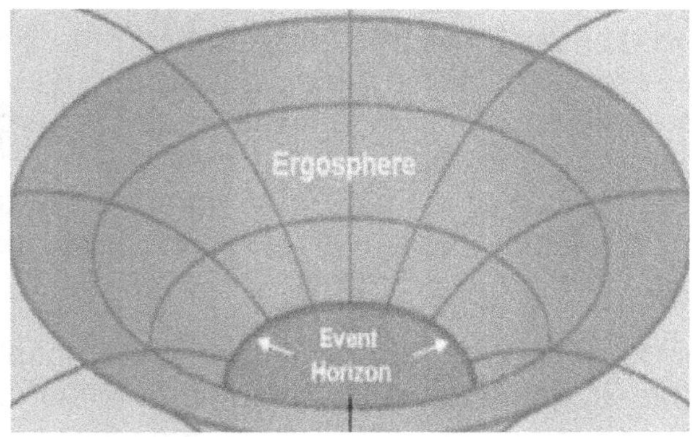

The ergosphere is relatively quite large and it contains atoms in the form of things that are being continuously or sporadically swallowed by the black hole. Things such as cosmic dust, rocks, asteroids, gas, and even stars and planets can be falling into the ergosphere of a black hole, and that is why it appears as a black hole in space.

So there is a continuous or sporadic stream of objects falling into the ergosphere of a black hole, and from there continuing down beyond the event horizon. The ergosphere spins very fast because it is part of the spinning black hole as a whole. This spinning sends any debris in the ergosphere in a downward direction past the event horizon, and into the black hole proper.

Any light rays that happen to go into a black hole will be destroyed by hitting the inner side of the ergosphere or going beyond the event horizon. But also, some of the light rays will be absorbed (i.e. refracted) by atoms in objects that are still in their downward journey of the ergosphere. Once absorbed by the electrons in the atoms of ergosphere-objects, such light is destroyed and new photons will be emitted by the electrons in the normal way. These newly emitted photons will radiate out in all directions as is normal for light.

In doing so, some of the emitted photons in the ergosphere will go in straight lines within the black hole and be destroyed, while others will radiate in straight lines out of the ergosphere. Such photons can do this because they are not subject to gravity and they can only travel in straight lines at the speed of light, which is faster than anything moving or spinning inside a black hole.

Such light does not spin around inside the ergosphere with everything else, going down towards the event horizon. Why not? Because light can never bend so it cannot spin around inside the ergosphere. And furthermore to do so the light would have to slow down to the ergosphere's spinning speed, which it cannot do.

So some of the incident light emitted out of debris in the ergosphere will shoot out of the ergosphere. The word 'inadvertently' is used bearing in mind that light is not subject to gravity, not even the gravity inside the ergosphere of a black hole. These newly emitted photons that come out of the ergosphere are referred to as Hawking Radiation:

"Hawking radiation is made almost exclusively of photons, which get emitted from a large region [called ergosphere] *outside the event horizon, not right at the surface"* (source: Ethan Siegel, Yes - Stephen Hawking Lied To Us All About How Black Holes Decay, Forbes.com).

The physicist Steven Hawking was incorrect to say that light escapes from black holes as a result of Einsteinian space curvature, but he was the first to formulate a theory showing that light did indeed 'escape' from black holes, and hence the name 'Hawking Radiation.'

Hawking was also incorrect to propose in 1974 that black holes eventually evaporate by losing what's now known as Hawking Radiation. This is so because any loss of radiation emanating from material that has entered the ergosphere has no impact on the total intrinsic energy of a black hole below the event horizon.

But, of course, any radiation escaping up and over the event horizon into the ergosphere and then out of the ergosphere would indeed reduce the overall intrinsic energy of a black hole. Such loss of energy would in total need to be greater than the total rate of energy going into a black hole. Hence, the so-called evaporation rate of a black hole will depend on this ratio of incoming and outgoing energy, assuming that black holes do indeed lose intrinsic energy.

Note: All Hawking Radiation is incident light. Therefore at some future date it should be possible for scientists to receive such light for spectroscopy analysis. This would allow a virtual video camera (see the section 'Virtual Video Camera') to be 'put' inside the ergosphere of a black hole to reveal much interesting information. Currently, any Hawking Radiation that is out there is many orders of magnitude below the current best telescopes' detecting ability.

Staying with black holes, in the real Universe that we inhabit the concept of infinity is not embraced in the FTOE. *"The entirety of physics' history has taught us that infinities do not exist in the natural world. When they arise in calculations, this is an indication that the theory is not completely correct and must be modified to eliminate them, replacing them with finite, albeit large, quantities"* (source: Andrey Feldman, New theory of gravity rethinks the Big Bang, Advance Science News, June 2023).

Hence, it is postulated that there is no so-called 'singularity' at the bottom of a black hole where matter is said to be 'pressed to infinity'. Rather, the matter/energy at the bottom of a black hole is simply very, very dense. It is speculated that the birth and death of a black hole may follow a bell curve. As the black hole is first formed it will have plenty to feed on in its vicinity and so will grow bigger and bigger. With time it will gradually exhaust things to feed on in its vicinity. This will then change the ratio of in/out energy, and gradually the black hole will become smaller and smaller and eventually peter out.

There is nothing at the bottom of a black hole except very, very dense matter

The James Webb telescope has revealed the presence of massive black holes in the early Universe, meaning that massive stars exploded as supernovae in the early Universe. Scientists are puzzled about this because it has been assumed that cosmic dust gradually came together to form opaque clumps of very cold gas and dust. And that when those clumps reach a critical mass, they collapse under their own gravity into a hot star. The conundrum is that this very long process of star formation is not thought to have occurred in the early Universe.

The *Final Theory Of Everything* tells us that star formation was entirely possible in the early Universe. UI gravity as postulated in this book will have pushed cosmic dust and gas together to form stars, thus readily leading to supernovae explosions and the formation of black holes in the early Universe.

Black holes, then, are simply a bunch of remnants that have coalesced from a supernova explosion, and they have come together to form a region of space with very dense matter, and hence very strong gravity. Scientists are urged to take into account the AE force and the nature of UI gravity as described in this book to better understand black holes.

Electromagnetism

In what follows we will look at the true nature of quantum gravity and how the gravity of small things is reconciled with the gravity of big things.

We live in a magnetic Universe. The Earth is magnetic. In fact, when we look out into space, all of the astrophysical objects that we see are embedded in magnetic fields. This is true not only of our neighbourhood of stars and planets, but also in the deep space between galaxies and galactic clusters. These fields are weak, typically much weaker than those of a refrigerator magnet, but they are dynamically significant in the sense that they have profound effects on the dynamics of the Universe.

What causes magnetism? Nearly everything in the Universe is made of atoms. Each atom has electrons, particles that carry electric charges. Spinning like tops, the electrons circle the nucleus, or core, of an atom. Their movement generates an electric current and causes each electron to act like a microscopic magnet. So all types of substances exhibit some degree of magnetism, however little.

Magnetism then is caused by movement (the movement of atoms). And since everything in the Universe is always moving, everything is magnetic to some degree. The technical nature of magnetism is well described in physics, so we won't repeat that here.

Electromagnetism combines the power of electricity and magnetism. Electromagnetism can be a temporary magnet: when an electric current is turned on it creates a magnetic field. When you turn off the electricity, the magnetism disappears.

Throughout the Universe, just about wherever there are stars and planets, superheated matter is also present. This superheated matter creates a soup of positively charged particles (ions) and negatively charged particles (electrons); in physics this is referred to as 'plasma'. This plasma in turn creates electricity and magnetism.

Dark matter

As mentioned, superheat comes from the remnants of the big bang which started the Universe billions of years ago. But this superheat is getting hotter. It is estimated that the Universe is about ten times hotter than it was about 10 billion light years ago. Why is this so?

The consensus view is that as the Universe evolves, gravity pulls together the dark matter and gas that is in space, forming galaxies and clusters of galaxies. The drag is violent—so violent that more and more gas is shocked and heated up. The problem with this view is that dark matter cannot be detected and may not even exist. Dark matter is merely a theory; it is inferred from the gravitational effect it seems to have on visible matter.

Furthermore, it is not suggested that the Universe as a whole is becoming hotter as this would violate the law of energy conservation. When it is said that the Universe is becoming hotter, this refers to heat radiating from all the matter that we can detect. Clearly, as more and more stars are formed, the overall heat of all matter will go up. But this increase in heat has to be balanced with the increase in areas of space devoid of heat.

Empty space has a temperature of absolute zero degrees kelvin. But in areas of space containing dispersed particles the temperature is not absolute zero. When dispersed particles come together to form stars, the temperature of space becomes higher (more concentrated) in the stars and colder (less concentrated) in the space that gave birth to the stars. So it all evens out. Dark matter is not needed to explain this phenomenon, and there is no mysterious unknown force that is making the Universe hotter.

Note: It is thought by some scientists that dark matter makes up most of the mass of galaxies and galaxy clusters, and is responsible for the way galaxies are organised on grand scales. Dark energy, meanwhile, is the name given to some kind of unknown mysterious influence driving the accelerated expansion of the Universe.

As mentioned in this book the accelerated expansion of the Universe began at the same moment that the Big Bang occurred. The new Universe needed to be created into a space where it could exist, so the Big Bang started creating that space at the same moment that the Big Bang occurred.

So if it is postulated that so-called 'dark energy' is responsible for driving the accelerated expansion of the Universe, a paradox arises: if dark energy is responsible for the accelerated expansion of the Universe, this means dark energy had to exist *before* the Big Bang so as to enable the Big Bang to expand into a space where it could exist.

Thus, if it is argued that dark energy is what creates the expanding space of the Universe, it would mean that the Big Bang would first have to create dark energy and then the dark energy would then create expanding space in which the big bang could be deployed - clearly absurd. And last but not least, what is the nature of this mysterious all-pervading dark energy? The existence of both dark matter and dark energy is very doubtful.

"A simple test suggests that dark matter does not in fact exist. If it did, we would expect lighter galaxies orbiting heavier ones to be slowed down by dark matter particles, but we detect no such slow-down. A host of other observational tests support the conclusion: dark matter is not there" (source: Pavel Kroupa, head of the Stellar Populations and Dynamics research group, professor of astrophysics at University of Bonn, Germany, July 2022).

*

"Spiral galaxies often have 'bars' in their central regions that rotate over time. If galaxies were embedded in massive halos of dark matter, their bars would slow down. However, most, if not all, observed galaxies have bars that are fast. This falsifies the standard cosmological model of dark matter with very high confidence. Also, it is suggested that the 'halos' around galaxies are caused by dark matter that provides gravity to the matter around it, but were not affected by the gravitational pull of the normal matter - this doesn't reflect reality. It is clear that dark matter halos around galaxies do not reliably explain their properties". (source: abridged extract from Dark Matter May Not Exist by Indranil Banik, postdoctoral research fellow of astrophysics, University of St Andrews July 10, 2022).

Note: The topic of Dark Matter is also discussed later in the section 'The Myth of Dark Matter'.

Coming back to why the Universe (or parts of the Universe) may be heating up, a likely explanation is that as the Universe expands it moves galaxies apart from each other, and this movement is what causes superheated gas. When you make something move you create heat. Thus, movement excites subatomic particles and makes them vibrate and rotate more quickly, and this creates heat.

The expansion effect of the Universe creates a continuous and accelerating movement of subatomic particles. This in turn creates superheat, which in turn creates plasma, which in turn creates electromagnetism. So movement (i.e. the expansion effect) is at the heart of electromagnetism in the Universe.

To summarise, the entire Universe is magnetic. From stars to galaxies to intergalactic space, magnetic fields thread the cosmos. Our home galaxy, the Milky Way, hosts a magnetic field that helps to shape the interstellar medium: the 'stuff between the stars' out of which new stars are born. This electromagnetism is born from the movement caused by the accelerating expansion of the Universe.

Electromagnetism in subatomic particles

Everything in the Universe that exists has mass and is therefore subject to gravity. Even atoms have mass and are subject to gravity. All atoms have a central mass (its nucleus) and orbiting particles. The nucleus of an atom is mostly made of protons. And the atom's orbiting particles mostly consist of electrons. We use the word 'mostly' to keep things simple as other types of subatomic particles can also be involved.

Thus, both the nucleus of atoms (i.e. the protons) and the orbiting electrons have mass. If this were not so, atoms would fall apart and the Universe as we know it would not exist.

Nothing in the Universe is massless. If it is massless it does not exist. When scientists say, for example, that gluons are massless, this is incorrect. Gluons have kinetic energy by virtue of their movement. And kinetic energy results in kinetic mass. Gluons and photons may also have 'proper' mass, but given current technology the amount of proper mass is too little to be detected by today's science.

Thus, it is generally accepted in physics that atoms and subatomic particles have mass. And given that they have mass, it means they will also be subject to gravity, a kind of gravity known as 'quantum gravity', which brings us to electromagnetism.

The electromagnetic force is responsible for many of the chemical and physical phenomena observed in daily life. The electromagnetic attraction between atomic nuclei and their electrons holds atoms together. Electric forces also allow different atoms to combine into molecules, including macromolecules such as proteins that form the basis of life.

Let us remember that everything in the Universe is moving and following its own little dance. And all matter is made up of elementary particles which have a spin. This spin (i.e. movement) creates electrically charged particles, and this in turn creates magnetic fields around all matter, including planets and stars. On a cosmic scale, magnetism has an enormous influence on the structure of the Universe.

There is no mystery to the nature of electrically charged particles. They are simply very, very tiny bits of matter (i.e. particles) full of electrons. These electrons make the particles be 'electrically charged'. And what are electrons? They are tiny particles that form part of an atom. The point here is that magnetism is not some kind of mysterious, little understood force.

Put very simply, magnetism is a group of atoms moving and bumping into each other, and when this happens, the electrons in those atoms get 'kicked around'. Some of those electrons move from one atom into another creating a magnetic force of attraction. And some of those electrons get repelled by atoms that already have enough electrons, creating a magnetic repelling force. The magnet in your hand and the nearby nails are made of atoms, and when they get near each other the electrons in the atoms of the magnet and in the nails bump into each other, causing magnetism. It is the non-stop movement of the atoms that causes magnetism.

That piece of paper or pen on your desk is made of moving atoms, and hence the paper and pen have a microscopic amount of electromagnetism. This does not mean that all materials are 'obviously magnetic'. For example, so-called non-magnetic substances include plastic, rubber, aluminium, lead, brass and others. But even these non-magnetic materials contain electromagnetism, however little.

There are three forces at work inside the atom: the electromagnetic force, the Strong nuclear force, and the Weak nuclear force. The electromagnetic force is a force of attraction or repulsion between all electrically charged particles. One way or another, electromagnetism affects all types of subatomic particles.

On a macro scale planetary and stellar orbits are gravitationally bound, but on a subatomic-scale orbits are electrostatically bound. Why is this? The most likely explanation is because the electrostatic force inside the atom's nucleus is always repulsive. Put simply, if the nucleus becomes bigger it becomes unstable and decays.

Here is how electromagnetism ensures that atoms stay together to form bigger objects:

"The electromagnetic force binds negatively charged electrons to positively charged atomic nuclei and gives rise to the bonding between atoms to form matter in bulk. Gravity and

electromagnetism are well-known at the macroscopic level" (source: Britannica.com).

Note: For the technically minded, in reality electrons and other subatomic particles don't actually spin in the classical sense of the word, like the spinning or rotation of a planet. So although electrons do not spin like little balls going around a bigger central ball, they do have a property called spin. This spin of subatomic particles is based on intrinsic angular momentum, whereas the spin of objects in space, such as stars and planets, is based on orbital angular momentum. Intrinsic spin is angular momentum (i.e. movement) caused by magnetism.

A crucial point here is that electromagnetism keeps subatomic particles spinning in orbit or grouped together with other atoms. And that is made possible by the AE force (accelerating expansion force of the Universe). Here is a step-by-step summary of how the AE force gives gravity to subatomic particles (points numbered 1 to 7):

1. AE force. The AE force exerts a force of inertia on everything that exists, including subatomic particles. It does this as a consequence of the continual accelerating expansion of the Universe. Inertia is a property possessed by anything with mass. We will remember that everything that exists has mass, and equally everything that exists is always moving. When a mass is moving at an accelerated rate it has inertia.

But when it comes to subatomic particles, the mentioned force of inertia is too weak to be accountable for quantum gravity. Nevertheless, subatomic particles do indeed have gravity. This is fully accepted by science generally, and such gravity is very evident from the force it takes to accelerate them in particle accelerators or electron beam instruments.

So how exactly does the AE force give gravity to subatomic

particles? It does this by expanding the whole Universe and in so doing making subatomic particles (and everything else in the Universe) move. This movement (or motion) creates kinetic energy, and this in turn creates electromagnetism, leading to quantum gravity.

To clarify further, the AE force exerts a pushing force on subatomic particles by trying to put more space between them as a result of cosmic expansion. But if subatomic particles are close enough to each other, their quantum gravity will prevail (they will not fly apart).

"It became apparent very quickly — as early as the 1930s — that there are no two ways about it: the Universe is, in fact, expanding. The fact that the redshift of an object matched up to the distance relation and the observed expansion rate as well as it did, no matter how far away an object was, helped confirm that" (source: Ethan Siegel, How Do We Know Space Is Expanding? Forbes.com). **Note:** more on redshift in the section 'The Doppler Effect'.

2. Kinetic energy. The AE force makes everything move in the Universe, including subatomic particles. When things move, the movement itself creates kinetic energy.

"In physics, the kinetic energy of an object is the energy that it possesses due to its motion. Kinetic energy is a property of a moving object or particle and depends not only on its motion but also on its mass" (source: Britannica.com).

Anything that moves will have mass and kinetic energy, however little. And it is well-known that kinetic energy creates electrical currents. These electrical currents in turn create electromagnetism. This is how electromagnetism is created in atoms.

To reiterate this crucial point, the accelerating expansion of the Universe creates space around objects, either pushing them apart, or pushing them together depending on their proximity to each other. This pushing force makes everything that exists move, thus creating kinetic energy and electromagnetism in subatomic particles.

"Objects in motion are examples of kinetic energy. Charged particles, such as electrons and protons, create electromagnetic fields when they move, and these fields transport the type of energy we call electromagnetic radiation, or light" (source: Tour of the Electromagnetic Spectrum, Nasa Science, December 2022).

3. Micro Electromagnetism. The electromagnetism of the atom is what holds subatomic particles together through attraction of particles with opposite charges. Specifically, the electromagnetic force holds the electrons around the nucleus of the atom. These are held together because electrons have a negative charge, and the nucleus has a positive charge. Electromagnetism works through the strong and weak fundamental forces ensuring that stable atoms can be formed and that chemistry, including the chemistry of life, can happen.

Technically, electromagnetism generates quantum gravity by making it possible for particles to be 'passed around'. For example, with the Strong force, quarks pass particles called gluons to other quarks, making such quarks stick together to form neutrons and protons.

And similarly neutrons and protons stick together to form individual atoms in a molecule. This 'passing around' of particles is made possible through electromagnetism, and the 'sticking together' is what we see as quantum gravity. Electromagnetism is at the heart of all this by enabling electrically positive protons to be held together with electrically neutral neutrons, thus making it possible for whole, stable atoms to be formed.

To clarify further, electromagnetism produces quantum gravity because electromagnetic fields themselves carry energy (and momentum and stresses). The energy density carried by an electromagnetic field can be computed by adding the square of the electric field intensity to the square of the magnetic field intensity. As another example, a beam of light (produced from, say, a laser) consists of an electromagnetic field, and it will exert a force on charged particles. Thus the electromagnetic field carries momentum. Because an electromagnetic field contains energy, momentum, and so on, it will produce a gravitational field of its own. This gravitational field (i.e. quantum gravity) is in addition to that produced by the matter of the charge or magnet.

The reason that quantum gravity acts through the mentioned three fundamental forces (rather than 'relying' on inertial gravity) is because the maximum universal size of an atom anywhere in the Universe is limited (it is simply the way the Universe is made, it is what it is).

Technically, the maximum size of an atom is limited because of the Strong nuclear force. The Strong force holds the nucleus of the atom together, but it has a very short range. Beyond a certain limit the Strong force will collapse and the atom will fall apart, thus limiting the maximum size of an atom in all parts of the Universe.

4. The Strong & Weak forces. As with electromagnetism, the Strong and Weak nuclear forces are also born out of movement. And their movement determines their ultimate destiny. Put simply, certain movements in subatomic particles cause those particles to change into other types of particles through their interaction, including the creation of Strong and Weak force particles. Technically, the Strong force changes quarks into protons, neutrons, and other hadron particles, and binds them into the nucleus of an atom. The weak force changes protons into neutrons and vice versa.

The electromagnetic force keeps the electrons attached to the atom. The Strong nuclear force keeps the protons and neutrons together in the nucleus. And the Weak nuclear force controls how (and if) the atom decays. So these three forces work 'in concert' to ensure that atoms stay together, but without touching each other, thus making it possible for the Universe to exist as we know it.

Atoms don't touch each other because electrons are whizzing around the outside of the atom, and they are the first things to interact with anything coming close to an atom. Since electrons in all atoms have the same charge, they repel one another, thus preventing atoms from touching each other.

As the Universe expands at an accelerating rate it puts more space between things. But if things are close enough (such as the atoms in our body, or the planets in our solar system) the 'closeness' trumps the pressure to separate exerted by the expansion effect.

Thus, even though the expansion effect is not enough to physically separate subatomic particles, it is nevertheless enough to exert a resistance to separate. It is this resistance to separation that excites subatomic particles, making them move more quickly. And when they move more quickly there is more electromagnetism, stronger quantum gravity and a resistance to separation.

So this resistance to separation arises from the AE force that causes subatomic particles to keep moving. This constant movement is what causes the Weak and Strong forces to come into existence and affect subatomic particles, each in their own different way.

5. Einsteinian gravity. The interaction (i.e. movement) between the three fundamental forces (electromagnetism, Strong force & Weak force) determines how subatomic particles behave; this interaction is what is known as 'quantum gravity'. The current consensus among physicists is that quantum gravity *"attempts to describe gravity according to the principles of quantum mechanics based on the three fundamental forces of electromagnetism, strong and weak forces. And that the fourth force, gravity, is based on the theory of relativity* [spacetime gravity] *which is formulated within an entirely different framework of classical physics. Quantum gravity is incompatible with* [spacetime] *gravity"* (source: Quantum Gravity, Wikipedia.org**).**

As postulated in this book, Einsteinian gravity, based on some kind of 'curvature of space', is considered to be baseless at both the macro and micro level, and it certainly has no part to play in quantum gravity. If you accept the concept of spacetime gravity you also have to accept the concept of two fundamentally different types of gravity, one for the big and one for the small.

If you accept the FTOE (*Final Theory Of Everything*) as postulated in this book you are faced with just one fundamental cause of gravity, common to both the big and the small: the AE force and the movement it creates.

6. UI gravity. Up until the present day it has not been known how subatomic particles acquire gravity, although a variety of theories have been proposed. But now we know: subatomic particles acquire their gravity from the AE force, as explained in this book. Think of it like this, the AE force gives UI gravity (inertial gravity) to everything that exists, including subatomic particles. But in the case of subatomic particles, UI gravity is not enough (it's too weak), so the AE force comes to the rescue by making subatomic particles move.

"Gravity is a real weakling, it is 10^{40} times weaker than the electromagnetic force that holds atoms together" (source: Why is gravity so weak? New Scientist.com).

This movement of subatomic particles provides the infrastructure for the three forces (electromagnetism, Strong & Weak forces) to come into play and provide quantum gravity. So on a micro scale the AE force provides the infrastructure upon which electromagnetism plus the Strong & Weak forces can come into existence and provide the kind of gravity we observe in subatomic particles.

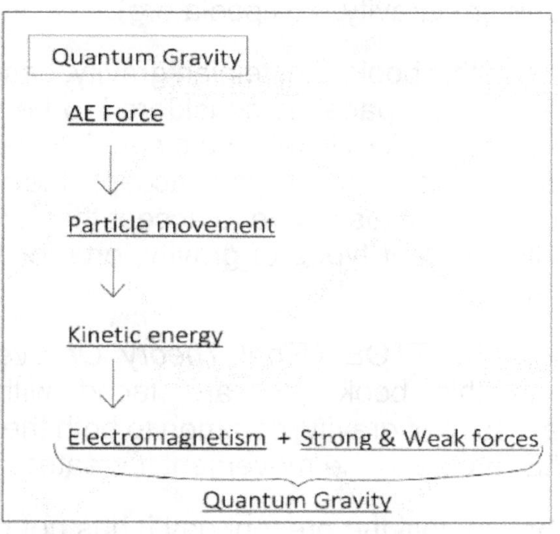

To reiterate, the AE force creates movements in subatomic particles. These movements create kinetic energy and electromagnetism. This in turn creates the Strong and Weak fundamental forces that help to keep atoms together to form larger objects. Thus, the AE force works through movement to give rise to the three fundamental forces of subatomic particles, and in this way endow subatomic particles with quantum gravity.

7. Joining the dots (of subatomic particles)

Whole books have been written on how subatomic particles behave, how they interact, how they come about, change, and decay. We won't attempt to emulate such learned and comprehensive information. The focus here is on the link between the accelerating expansion of the Universe and how this AE force leads to quantum gravity, a topic which has been covered on and off throughout this book. What follows is a summary of this direct link between the AE force and quantum gravity.

The behaviour of matter at the molecular scale including its density is determined by the balance between the electromagnetic force and the force generated by the exchange of momentum carried by the electrons themselves.

Photons exert a different pressure on electrons than protons, and also they scatter off electrons more often. It has been found that the differences in the movements of electrons and protons will generate rotating electric current and magnetic fields. In short, quantum gravity is based on electromagnetism, and electromagnetism is based on movement created by the AE force.

It could be argued that electromagnetism is created by rotation rather than by mere movement. It is thought by physicists that almost everything in the Universe spins. Particles spin around in atoms, planets rotate on their axis, stars rotate around galaxies, and whole galaxies spin in great spiral structures. **Note:** it is not known whether the whole Universe is rotating, but current consensus is that it is probably not doing so.

This rotation or spinning movement is indeed at the heart of existence because such movement creates kinetic energy, electric currents, magnetic fields, electromagnetism, and the Strong and Weak forces. So it is argued that rotation is at the heart of existence. But what causes such universal rotation?

The standard model of physics gives the following answer:

Quote

There is no force that causes objects such as planets to rotate. Most of the rotation comes about from the conservation of angular momentum. Due to conservation of angular momentum, if the radius of the orbit decreases, then its angular velocity must increase (as the mass is constant). All planetary and stellar systems are born from the collapse of dense interstellar clouds. As it collapses, it coalesces ever so slightly into an angular momentum. This creates a very slight rotation that the cloud has in the beginning and is increased dramatically when the collapse takes place. So the rotations that we see today are due to this original angular momentum arising from interstellar clouds of particles.

Unquote

In the above quotation it is incorrect to say that "*there is no force that causes objects to rotate*". There must be a force that makes objects coalesce and rotate if they are close enough to be affected by each other's gravity. Otherwise, how do you explain that clouds of objects are able to 'collapse' into stars and planets? A force must exist that is responsible for making the objects in the cloud move together, otherwise no collapse would occur. The cloud would continue as a cloud. How can the "*radius of the orbit decrease*" (so as to create angular momentum) if there is no coalescing force to make the radius decrease?

Coming back to joining the dots of subatomic particles and quantum gravity we have three important points:

1. Subatomic particles. All matter in the Universe started as subatomic particles that gradually coalesced as a result of quantum gravity. *"In the first moments after the Big Bang, the Universe was extremely hot and dense. As the Universe cooled, conditions became just right to give rise to the building blocks of matter – the quarks and electrons of which we are all made"* (source: The Early Universe, European Organization for Nuclear Research).

From reading this book you will know that the force that causes objects to rotate is the AE force. The AE force keeps everything moving or rotating in the Universe. When particles in a dust cloud move, they bump into each other causing tiny amounts of spin, kinetic energy, and electromagnetism. This in turn leads to a coalescence of particles and angular momentum.

On a small scale, when dust particles move they take on a positive or negative static electric charge due to contact with other dust particles. This creates magnetism, and hence coalescing. On a larger scale than specs of dust, such as rocks, the AE force will exert a pushing (separating) force arising from cosmic expansion. But if such rocks are close to other rocks, such as in a cosmic cloud of dust, the pushing force of inertial gravity will prevent the rocks from separating.

2. Angular Momentum. As the objects (dust and rocks) gradually come together, some of them will spin a little as a result of bumping into each other. Various combinations of spin, magnetism and inertial gravity will ensue, making any spinning motion go faster.

The conservation of angular momentum can explain why objects spin in space. If an object has even a little bit of spin, such spin becomes faster due to the effect of inertial gravity. One can see an example of this concept by watching an ice skater. When an ice skater starts to spin, the skater will have the arms stretched outward. At this point, the skater rotates slowly. However, when the skater brings the arms in, the spinning becomes faster and faster.

3. Going from a cloud to a disk. When the skater's arms come down you in effect reduce the body's desire to shoot off in a straight line (you 'conserve angular momentum'). This then allows the skater's kinetic energy to be focused on spinning. Thus when rocks and dust particles bump into each other in space, this creates a little bit of spinning which in turn creates a little bit of kinetic energy, and hence a little bit of gravity. This in turn attracts other nearby rocks and dust particles to coalesce, and gradually the object grows in size and mass, eventually becoming a star or planet.

As spinning clouds travel through space, they tug on other objects with their gravity. The particles inside the cloud move around, and as they hit each other, they cancel each other's prior movements. Everything is still spinning. Over time, as the cloud spins it becomes thinner akin to a pancake, creating a disk. The planets and stars in a galaxy all formed out of the same spinning disk of gas. That is why they spin and travel together in the same direction.

As the Universe expands at an accelerating rate it exerts a pushing force on everything that exists (whether it be pushing to separate or pushing to coalesce). So in the case of star formation, the AE force tries to push apart all the objects in the cloud that is destined to become a star. This pushing apart force puts inertial gravity around each of the objects in the cloud. But as such objects are close enough to be affected by each other's UI gravity, the objects coalesce instead of separating, as explained in the section '*The Coalescing Force*'.

As the objects coalesce, the angular momentum arising from the coalescing eventually results in the rotation of the star on its axis as it is formed. A similar explanation can be given for the cause of rotation of planets, solar systems and whole galaxies.

When it comes to the rotation of subatomic particles, the explanation is similar.

According to CERN, which is the European Council for Nuclear Research, atoms were created about 13.8 billion years ago in the first few minutes after the Big Bang. The new Universe quickly expanded, creating the conditions for electrons and quarks (the smaller particles that make up protons and neutrons) to form. Millionths of a second later, quarks aggregated to form protons and neutrons, which combined to form the nuclei of atoms.

So during the moment of the Big Bang an enormous amount of energy was released, and because matter and energy are different forms of the same thing, some of that energy manifested as fundamental particles, to provide the building blocks of atoms.

As the Universe began to expand, things eventually spread out and slowed down to the point where clouds of quarks and electrons were formed. The quarks coalesced into protons and neutrons, and together with electrons, atoms were formed. This all took place a few hundred thousand years after the moment of the big bang, and this marked the formation of stable atoms.

What made the subatomic particles coalesce together to form atoms? Answer: the AE force. The moment the Big Bang occurred the accelerating expansion of the Universe also began. Without the AE force there would have been no space in which the new Universe could be manifested.

We have mentioned that as the Universe expands at an accelerating rate it exerts a pushing force on everything that exists. So in the case of the formation of atoms, the AE force tries to push apart all the subatomic particles in the cloud that is destined to become a bunch of atoms. This pushing apart force makes the subatomic particles in the cloud move more quickly, which in turn creates kinetic energy, electrical currents, electromagnetism, the Strong & Weak forces and quantum gravity.

As quantum gravity takes hold in the 'cloud' of subatomic particles, they coalesce because they are close enough to be affected by each other's newly formed gravity. As they coalesce, the angular momentum arising from the coalescing eventually results in the spinning of the subatomic particles that make up each atom.

Technically, there are two things that contribute to angular momentum: spin, which is the intrinsic angular momentum inherent to any fundamental particle, and orbital angular momentum, which is what you have from two or more fundamental particles that make up a composite particle.

We conclude then that the rotation (or spinning) of just about everything in the Universe is indeed at the heart of creation by virtue of the fact that such rotation gives existence to electromagnetism, and this in turn gives existence to atoms and subatomic particles and the Universe as we know it. But the accelerating expansion of the Universe is the force behind the mentioned rotation and spinning of objects. The AE force is at the heart of creation.

As mentioned, the AE force is the common link between macro and micro gravity, and there is growing evidence that the same gravity we experience everyday also affects subatomic particles. For example, in February 2024 a team of scientists at the University of Southampton, UK, has pushed the boundaries of physics by measuring gravity's effects on subatomic particles. Einstein was always sceptical about the practicality of testing gravity's effects at a subatomic level, but the Southampton University experiment proves it is indeed possible. And it points towards validating a single force of gravity common to the big and the small.

By demonstrating experimentally that gravitational forces can indeed be detected and measured on the smallest of scales, scientists have opened up new avenues for research that can lead to a fuller understanding of quantum gravity that brings together a grand unification theory as postulated in this book.

Tim Fuchs, a scientist at the University of Southampton, said: *"By understanding quantum gravity, we could solve some of the mysteries of our Universe — like how it began, what happens inside black holes, or uniting all forces into one big theory"* (source: article by Robert Lea, Space.com, February 2024).

The take home message: The quantum gravity of subatomic particles arises from the AE force, which in turn gives birth to electromagnetism. This electromagnetism brings into being the Strong and Weak forces making it possible for atoms to be formed and for quantum gravity to exist. The rotation or spinning of just about everything in the Universe is caused by the AE force, a fifth fundamental force that is at the heart of creation.

The Mathematics of the FTOE

This section explores the mathematics behind the *Final Theory Of Everything*. For those who may not be mathematically inclined please feel free to skip this section. The mathematics for the concept of gravity and the *Final Theory Of Everything*, as postulated in this book, is well-established and does not involve any kind of Einsteinian relativity or the re-inventing of the mathematical wheel.

If you find that the mathematics of physics is usually too complicated, you are not alone. There is a tendency to overcomplicate the mathematics of astronomy and cosmology among professional physicists and professors. This happens mainly because the primary concern of professional physicists is to have credibility in their profession. So there is a strong tendency to publish journal articles (and sometimes give incomprehensible lectures!) in which they use very rigorous, jargon-heavy, over-complicated mathematics. Many find it difficult to explain their concepts as simply as possible for fear of losing credibility.

The saying that *"if you can't explain it simply, you don't understand it"* has merit. But of course many concepts in physics do indeed require a 'higher level' of mathematics that is beyond the grasp of most people. But such concepts can usually be simplified so as to give a basic understanding to non-mathematicians.

However complex the mathematics, the concepts can always be simplified. For example, in 1965 the physicist Richard Feynman won the Nobel prize for his contributions to the development of quantum electrodynamics, a mathematically complex area of physics. Feynman developed a greatly simplified pictorial representation scheme for the mathematical expressions that described the behaviour of subatomic particles (such pictorial representations later became known as the Feynman diagrams).

In this book we talk about UI gravity (Universal Inertial gravity). We say that UI gravity is caused by the AE force (Accelerating Expansion force of the Universe). The AE force exerts a force of resistance to changes in velocity by virtue of pushing objects apart, or pushing objects together, depending on their proximity to each other. This pushing force, arising from cosmological expansion, creates inertia and hence gravity.

In physics, this mentioned inertial gravity is known as 'inertial mass'. Inertial gravity and inertial mass are two different names for the same thing. Their force is measured by measuring an object's resistance to changes in velocity.

This is expressed in the equation F = ma, where force is in newtons (N), mass is in kilograms (kg), and acceleration is in metres per second squared (m/s^2). This equation shows that mass has inertia; the larger the mass the greater the amount of force (F) needed to cause a change in motion (a).

As an example, consider an object of unknown mass. A force of 2N is applied on the mass, causing it to accelerate at 5 metres per second squared. The equation F = ma can be rearranged for m: F/a = m. Plugging in 2 N for (F) and 5 metres per second squared for (a) yields m = 0.4 kg. Now consider the same force applied to a different mass, resulting in an acceleration of 1 metre per second squared. Plug in 2 N for (F) and 1 m/s^2 for (a). Solving for mass shows that the object has a mass of 2 kg.

The mathematics for inertial gravity is well-established and applicable to all types of inertia, whether it be acceleration in a car, a rocket in space, or the inertia caused by the AE force to all parts of the Universe.

The mathematics of cosmological expansion

The accelerating expansion of the Universe (the AE force) can be calculated in various ways. The method most favoured by physicists is the 'Hubble flow' method. It is described by the equation v = H0D, with H0 being the constant of proportionality (the Hubble constant) between the "proper distance" D to a galaxy, which can change over time, unlike the comoving distance, and its speed of separation v, i.e. the derivative of proper distance with respect to the cosmological time coordinate.

The Hubble constant is most frequently quoted in (km/s)/Mpc, thus giving the speed in km/s of a galaxy 1 megaparsec (3.09×1019 km) away, and its value is about 70 (km/s)/Mpc. However, crossing out units reveals that H0 is a unit of frequency (SI unit: s−1) and the reciprocal of H0 is known as the Hubble time. The Hubble constant can also be interpreted as the relative rate of expansion. In the form H0 = 7%/Gyr, it means that at the current rate of expansion it takes a billion years for an unbound structure to grow by 7%.

Here is an answer to the question: If gravity is the same as acceleration, then does the accelerating expansion of the Universe cause gravity? The answer from cosmologist Jen Jamison is as follows:

Yes - Cosmological expansion is the cause of isotropic spatial acceleration - not in the sense that expansion causes masses to increase in volume but because space increases in volume. The global isotropic acceleration of space created by cosmological expansion does not disassociate matter, but the effect of isotropic spatial acceleration acting upon a massive body creates an inertial reactionary force. You can calculate it very easily - take the earth as an example, and use Newton's second law to calculate the force created upon the mass of the earth by the acceleration of space (c^2)/R. Then divide that force by the surface area of the earth. The earth's mass M is roughly 5.98 x 10^24 kg and its radius 'r' is 6.37 x 10^6 metres. So the total acceleration is [(c^2)/R] and the total force is [(c^2)/R] x M. The force per unit area is:

F/A = [(c^2)/R]M/4(pi)r^2

You will get your answer in units of ntn per square metre. To get it in units of ntn/kg, the cosmological acceleration factor (c^2/R) needs to be converted to volumetric acceleration per square metre. So multiply (c^2/R) by (metres^ 2/kgm). Source: Jen Jamison, Cosmologist & pioneer of inertial guidance systems, USA.

It is interesting to note that these numbers show a strong correlation between the expansion of the Universe and the force of gravity, adding fuel to the concept that gravity is borne from the AE force.

Consider that the rate of cosmological volumetric acceleration has precisely the correct rate to explain the local g fields of the masses. What we observe as the g field of a Mass M such as the earth is actually the inertial reaction of M to the 4(pi)G cosmological expansion field. One is normally used to seeing this in the form of Newton's law of inertia aka his 2nd law of motion:

F = Ma .... (1)

Where F is the force created by a mass, and M is the rate of acceleration with respect to the Universe. Put simply, gravity is the result of Mass multiplied by the Cosmological expansion factor 4(pi)G. How do we know 4(pi)G is the volumetric acceleration factor of the Hubble Constant? Because we derive G from Friedmann's equations or by dividing Newton's law of gravity by his 2nd law of motion equations.

Many scientific papers and articles have been published showing a variety of ways of calculating the accelerating rate of expansion of the Universe. We won't delve into this mathematically complex arena; suffice to say that whatever the precise rate of expansion, the AE force exists and is responsible for the creation of the Universe as we know it. The AE force makes everything move, thus allowing atoms and life to be formed, allowing objects to be endowed with UI gravity, and allowing subatomic particles to be endowed with quantum gravity.

The mathematics of the AE force

As explained in this book, the AE force (Accelerating Expansion force) of the Universe gives gravity to everything that exists. The force (strength) of such gravity is the same everywhere because the rate of acceleration (of cosmic expansion) is constant and universal. Clearly, the force of gravity will vary from object to object depending on its particular mass, location and acceleration at the time.

But in regard to the AE force, the following question arises: To what extent do objects in the Universe experience a reactionary force due to the isotropic acceleration of space? Put another way, what is the strength of gravity experienced by objects arising from the AE force (i.e. from the inertia caused by the AE force)?

We can use Newton's 2nd Law and our planet Earth to answer the question. The acceleration of space is c^2/R - that is the acceleration you would measure at any point on the Hubble surface.

Note: the 'Hubble surface' refers to a spherical region of the observable Universe in which objects recede from an observer due to the accelerating expansion of the Universe.

We do not see anything beyond this spherical region because it is theorised that objects recede from an observer at a rate greater than the speed of light due to the expansion of the Universe.

So F = Ma

Earth's mass is about 6 x 10^24 kg, and the Hubble scale R is about 10^26 metres.

So the total force of gravity is about [(9 x 10^16)/(1 x 10^26)] x 6 x 10^24 = 5 x 10^15 ntn.

But if you wanted to know the intensity (strength of gravity), you would need to divide it by the earth's surface area A. The earth's radius r is about 6.37 x 10^6 metres, so the Earth's surface area is:

A = 4(pi)(r)^2 = 4 x 3.14 x 40.5 x 10^12 = 509.64 x 10^12 square metres.

Therefore:

SOLUTION ONE. Intensity (strength of gravity on Earth) = [5 x 10^15]/[509 x 10^12] = 9.80 ntn/metre^2. This can also be expressed as 9.80 m/s^2. This result has been arrived at by invoking the AE force, i.e. by invoking the accelerating expansion of the Universe.

But there is another way to measure the strength of gravity on Earth without reference to the AE force. It goes like this:

SOLUTION TWO. Intensity (strength of gravity on Earth) is calculated as the speed of an object falling freely near the Earth's surface. As it falls it will increase by about 9.81 metres (32.2 ft) per second every second. This calculation ignores the effects of air resistance.

The precise strength of Earth's gravity varies depending on location. For example, the average equatorial value of gravity on Earth is 9.78 m/s^2. The nominal 'average' value at the Earth's surface, known as standard gravity is, by definition, 9.80665 m/s2 (about 32.1740 ft/s2).

So in **SOLUTION ONE**, the AE force gives us a strength of gravity on Earth as 9.80 m/s^2. In **SOLUTION TWO**, without taking into account the AE force, the strength of gravity is calculated to be 9.81 m/s^2.

The similar results from the above solutions one and two (9.80 m/s^2 and 9.81 m/s^2 respectively) is strong evidence for corroborating the Universal Inertial gravity emanating from the AE force.

The mathematics of the Gauss Law of Gravity

Carl Friedrich Gauss, a renowned German mathematician, said that only two things are required for inertial gravity: an accelerating volume G and an inertial mass M. The mass can be of any size, shape, temperature or density, and since mass and inertial gravity go hand in hand, it means the same kind of gravity applies to everything that exists. The resultant g* field is 3-D because G is an accelerating volume.

Inertial Mass M plays the same part in gravity as it does in Newton's second law - namely as an inert couch potato which opposes all acceleration - whether it be due the acceleration of the mass itself or to the volume of the Universe in which it is cast. In other words, gravity is the inertial reaction of mass to acceleration, and like all 2nd law forces, all gravity fields are proportional in strength to the mass from which they appear to emanate.

Gauss's law can be derived from Coulomb's law and depends on the inverse square proportionality which is also seen in the gravitational law formula. Only the proportionality constant is different. Therefore, Gauss's law is also applicable for gravitation (using the right constants).

In the context of inertial gravity when Gauss says that mass can be of any size, shape, temperature or density, Gauss is affirming that inertial gravity can apply to any kind of object in the Universe. And it should be noted that the Gauss Law of gravity is in keeping with Maxwell's and Newton's equations rather than Einstein's equations. Mathematicians are well aware that the mathematics of general relativity does not satisfy the Gauss Law of Gravity.

Gauss basically says that the total gravitational flux emanating from a sphere enclosing the Earth is 4πGM. So if we now divide this by the total surface of the sphere 4πR2, with R the radius of the Earth, we get: GMR2.

This gives the Gauss gravitational flux density. If you then calculate the numerical result for GMR2 you get a force of gravity equal to 9.81 m/s2 which agrees with the gravity value of the AE force and with other methods of calculating the force of gravity.

Here is a contribution from Jen Jamison, cosmologist, regarding Gauss and the law of gravity:

The simplest law for gravity is due to Gauss.

g = - 4(pi)GM

The least controversial way to express gravity is as an acceleration. More specifically, Gauss's law expresses the local acceleration field of a mass M in terms of the global acceleration field G of the Universe. But it gets even better.

g/M = -4(pi)[c^2]R/M'

which reads:

Local volumetric acceleration per unit mass M equals Hubble volumetric acceleration per unit mass M.

What is the significance of the accelerating expansion of the Universe? It makes us realise that the Universe is a self-creating cosmology. Gone are the days of the coasting Universe and the passive gravitationally slowed Universe. The Universe is actively participating in its own destiny. Gravity now makes sense because 4(pi)G[kg/metre^2] is equivalent to the accelerating expansion rate of space c^2/R. Source: Jen Jamison, Cosmologist and pioneer in inertial guidance systems.

The mathematics of gravity according to general relativity

Given the widespread, albeit mistaken, acceptance of general relativity, you have to ask yourself: *why are physics students still being taught gravity using Newton's laws rather than Einstein's general relativity?* The answer is that the mathematics of Newton's laws are so much simpler - it just requires simple algebra to express. Furthermore the mathematics of general relativity can only ever be an approximation because they don't take into account the force of inertia exerted by mass.

On the other hand Newton's law of gravity is based on mass and inertia and the mathematics is far, far simpler. Also it is accurate for all practical purposes. In Newton's day the phenomenon of kinetic energy was not known about. If the mathematics of kinetic energy is included in Newton's law of gravity, the results become virtually one hundred percent accurate.

Relativists are cagey when asked: *What exactly is the force of gravity on Earth, expressed mathematically, when calculated according to general relativity?*

Using relativity to calculate the force of gravity on Earth is said to require tensor calculus and four dimensional Riemannian geometry. It seems nigh impossible to find such mathematics laid out systematically step-by-step and culminating in a numerical result such as 9.8 m/s2.

This creates a serious conundrum for relativists: how do you measure gravity, based on relativity? To resolve this conundrum relativists resort to the Einstein Equivalence Principle. The EEP says that mathematically the force of inertial gravity is the same as the force of spacetime gravity.

Einstein developed the equations of his equivalence principle to show that the mathematics of inertial gravity and spacetime gravity are one and the same. He did this so as to show and prove the mathematical validity of spacetime gravity. But as explained in the section 'The False Equivalence Principle' there is no equivalence since spacetime gravity is baseless.

The mathematics of the Final Theory of Everything

UI (Universal Inertial) gravity as described in this book refers to the inertial gravity bestowed on all objects in the Universe. This inertial gravity arises from the AE (Accelerating Expansion) force of the Universe which acts to create more space around and between objects. But if objects are close enough to be affected by each other's inertial gravity, the AE force pushes (keeps) such objects together. In brief, gravity is the result of an inertial reaction created by cosmic expansion interacting with non-expanding matter.

So let's connect the dots: (1) AE force ➤ (2) cosmic expansion ➤ (3) Gauss inertial gravity ➤ (4) UI gravity ➤ (5) Quantum gravity ➤ (6) FTOE.

What are we missing? No *Final Theory Of Everything* would be complete without a gravity constant (abbreviated to 'G'). The gravity constant G describes the intrinsic strength of gravity, and can be used to calculate the gravitational force between two objects.

It was first defined by Isaac Newton in his Law of Universal Gravitation formulated in 1680. It is one of the fundamental constants of nature, with a value of (6.6743 ± 0.00015) x10^–11 m^3 kg^–1 s^–2. So the gravitational 'pull' between two objects is found by multiplying the mass of those two objects (m1 and m2) and G, and then dividing by the square of the distance between the two objects (F = [G x m1 x m2]/r^2). This explains why any object caught up in gravity will move at a constantly accelerating rate. It accelerates at a constant rate because of the gravity constant G.

But of course, the speed of the mentioned constant acceleration depends on the mass of the object. For example on Earth the constant acceleration is estimated to be 9.8 m/s^2. On the moon it is 1.62 m/s^2 because of the moon's smaller mass.

Why is there a gravity constant in the Universe? Where does it come from? It comes from the AE force, which in turn comes from the constant acceleration of cosmic expansion. This is how the gravity constant G comes into existence and fits into the FTOE.

The following four points are taken from comments made by Jen Jamison, Cosmologist & pioneer of inertial guidance systems, USA.:

Quote

1. Gravity is not the moderator of expansion — rather exponential expansion of the Universe is the cause of gravity. No Dark energy is required.

2. Inertial reaction and gravity are interdependent. The size and energy contained in a local gravity field increases as space expands. Orbits are stable not because G (gravity) is constant, but because the MG product is constant - for a circular orbit MG = (v^2)r. **Author's Note:** no orbits are 'stable' or circular, but on human time scales they appear to be so, and for all intents and purposes may be regarded as so.

3. The effective inertial modulus of free space is one kg/metre^2. The Universe acts as a virtual infinite plane area-density when opposing inertial reaction a la Newton's 2nd law. Not unrelated, is the fact that the observable Universe (Hubble bare mass taken as 1.5 x 10^53 kg) and effective Hubble area density (based upon a scale of 1.1 x 10^26 metres), also leads to a value of approximately one kg/m^2. A Gaussian transformation from 3-sphere to 2-sphere to infinite plane produces the same result.

4. The net zero energy Universe has the same solution as the empty Universe first discovered by Willem de Sitter in 1917. Space expands exponentially in both models - ergo as previously emphasised, no dark energy is required.

Unquote

The take home message: The mathematics for the *Final Theory Of Everything* already exist, it's just a matter of relating such mathematics to the FTOE. Among other things, the mathematics show that gravity is the result of inertia created by the accelerating expansion of the Universe. Thus gravity does not cause cosmic expansion, gravity is the effect of cosmic expansion. The AE force and UI gravity are well grounded in mathematics, thus giving validity to the *Final Theory Of Everything*.

FTOE (Final Theory of Everything)

This section discusses the fifth fundamental force of the Universe that brings together the various points in this book relating to the *Final Theory Of Everything,* and the common ground between inertial gravity and quantum gravity.

Einstein spent the last two decades of his life searching for a unified theory of gravity and electromagnetism and failed. The Grand Unified Theory proposed in 1974 by Sheldon Glashow and Howard Georgi to unify electromagnetism with the two nuclear forces also failed to be vindicated, including many of its more recent extensions.

The general theory of relativity is increasingly being shown to be wrong and there is a need for a theory that goes beyond general relativity into the quantum realm. Such a theory is often referred to as a 'theory of everything'. This book postulates such a theory, and is referred to as the *'Final Theory Of Everything'* (FTOE).

The physicist Werner Heisenberg once wrote that *"What we observe is not Nature itself but Nature exposed to our methods of questioning. What we can say about Nature depends on the precision and reach of our instruments dictating how 'far' we can see. Therefore, no theory that attempts to unify current knowledge can seriously be considered a 'final theory of everything' - we can never be sure that we aren't missing a huge piece of evidence".*

These wise words by Heisenberg are taken to heart, and the *'Final Theory Of Everything'* as postulated in this book is humbly put forward fully recognising that there will always be more to discover. As Richard Feynman said in relation to seeking scientific truth: *"If it turns out like an onion with millions of layers, and we're just sick and tired of looking at the layers, then that's the way it is".*

The FTOE fifth fundamental force

The FTOE describes a fifth fundamental force: the mentioned AE force. This Accelerating Expansion force of the Universe is what provides the framework for unifying all the four fundamental forces into a single mathematical framework. Here again is a list of the five fundamental forces of the Universe:

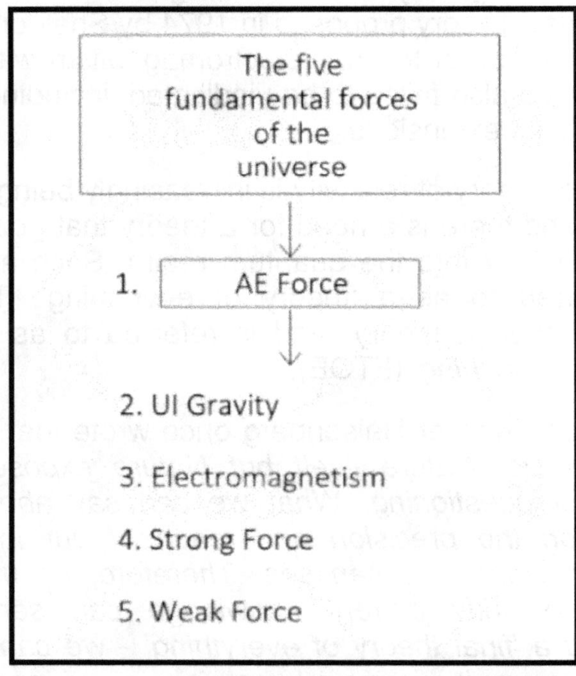

Hitherto, quantum gravity has not been able to explain how or why subatomic particles have gravity. But as discussed, now we know that quantum gravity arises from a series of events that go like this: AE force ➞ movement of particles ➞ kinetic energy ➞ electromagnetism ➞ Strong & Weak forces ➞ quantum gravity.

A key aspect of the FTOE is the fact that the Universe is expanding, but at an accelerated rate. It is this acceleration that gives rise to inertia, gravity and movement everywhere. Without the acceleration there would be no gravity, no formation of atoms and stars, and no Universe as we know it.

The expansion of the Universe has always been accelerating. Every second the total volume of the Universe becomes bigger and bigger. This is very much the consensus opinion among scientists today.

If the Universe were not expanding at an accelerated rate it would have stayed at the same size or indeed would have been contracting. But all the evidence is showing that the Universe is in fact expanding at a constant accelerating rate.

It is the acceleration itself that brings into existence inertia and gravity, and the formation of planets and stars in all parts of the Universe.

It is fully recognised that on a macro-scale, objects are gravitationally bound, and that on a subatomic-scale objects are electrostatically bound. For example, The force between an electron and a proton, which together make up a hydrogen atom, is about 36 orders of magnitude stronger than the inertial gravitational force acting between them.

We live in an electromagnetic Universe. Electromagnetism is everywhere because everything in the Universe is moving. And if it's moving it will have a degree of electromagnetism, however little.

The sequence of events leading to the FTOE goes like this:

(1) The AE force (accelerated expansion force of the Universe) pushes objects apart or pushes objects together, depending on their proximity to each other.

(2) This continual and accelerating 'pushing force' creates UI gravity in the macro Universe and quantum gravity in the micro Universe.

(3) The AE pushing force makes everything move and this creates kinetic energy, electrical currents and electromagnetism in all objects, big and small.

(4) In big objects inertial gravity trumps electromagnetic (quantum) gravity and we have UI gravity as described in this book.

(5) In small objects (subatomic particles) electromagnetic (quantum) gravity trumps inertial gravity and we have quantum gravity as described in this book.

(6) Both types of gravity (inertial gravity and quantum gravity) are born from the ***same*** accelerating expansion of the Universe - the AE force.

(7) If objects (big or small) are close enough they will coalesce over time, if not they will separate over time.

(8) Everything that exists is either moving apart or coalescing, there is no other type of movement in the Universe, and nothing can be stationary.

To be flippant for a moment, if you were to walk any which way, isn't this a different movement compared to the 'moving apart' or 'coalescing' movements? No, it is not different. Whichever way you walk, planet Earth (and everything on it) will be coalescing with the Milky Way galaxy as a whole, and simultaneously planet Earth (along with the Milky Way) will be separating from distant galaxies. It is the same if you fly to Mars: the rocket taking you to mars will be coalescing with everything in the Milky Way, and simultaneously the rocket will be separating from distant galaxies.

In his early life while developing his special theory of relativity, Einstein was adamant that there was no all-pervading ether in the Universe. At the end of his life, however, he changed his mind. He needed an ether to make his general theory of relativity work in his equations.

This 'general theory of relativity' resulted in many more unobserved and artificial mathematical consequences and ad hoc speculations. It also required the abandonment of Euclidean geometry and the invention of his mentioned bizarre new curved spacetime in order to mathematically make his theory of gravity work. Furthermore, the curvature of space between objects absolutely requires the existence of an ether, otherwise how could objects react to each other's presence so as to trigger space curvature?

Because of this conundrum, Einstein was forced to say that an ether existed (having denied the existence of an ether for many years), as otherwise he could not 'validate' his general theory of relativity. Albert Einstein stated the following in his talk to the University of Leiden on 5th May 1920:

"Recapitulating, we may say that according to the general theory of relativity space is endowed with physical qualities; in this sense, therefore, there exists an ether. According to the general theory of relativity, space without ether is unthinkable, for in such space there not only would be no propagation of light, but also no possibility of existence for standards of space and time (measuring-rods and clocks), nor therefore any space-time-intervals in the physical sense".

In short, Einstein was saying very clearly that gravity as postulated in his general theory of relativity would not work without an ether.

Many experiments since the 1920's show that an all-pervading ether in the Universe does not exist, thus invalidating Einstein's general theory of relativity and his theory of gravity based on space curvature.

As the acceptance of Einstein's Theory of Gravity gradually crumbles in today's world, scientists are looking for a new theory of gravity to explain how clusters of galaxies come together, what causes the Universe to expand, and how macro and micro gravity can be reconciled.

Note: For a more detailed and technical analysis of Einsteinian relativity and why it fails in today's world of physics a good starting point is a website called 'Millennium Relativity' (www.mrelativity.net).

If you have read this book up to this point you will now know that macro and micro gravity emanates from a common source: the AE force (Accelerating Expansion force). This AE force makes everything move and is the master fundamental force of the Universe that gives birth to the other four fundamental forces, namely gravity, electromagnetism, and the Strong & Weak forces. The AE force is the scaffolding upon which the other four forces rely.

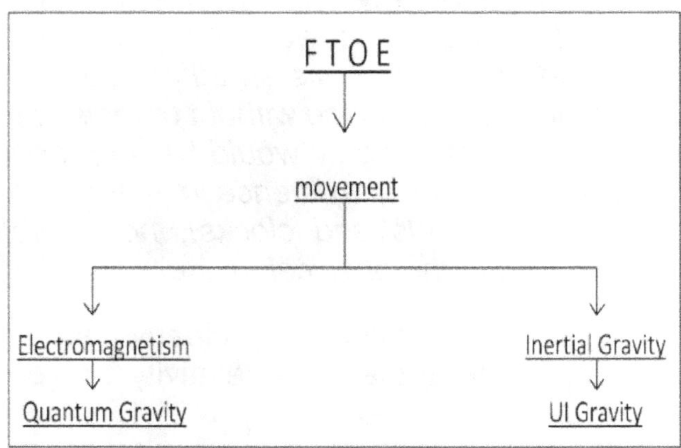

So we now have common ground between the macro

Universe of big things and the micro Universe of small things. By bringing together UI gravity (Universal Inertial gravity) and quantum gravity, under the umbrella of the AE force, we have a *Final Theory Of Everything* upon which to base our knowledge of cosmology and have a better understanding of the Universe.

The take home message: The standard model of cosmology is baseless because it depends on an ether and it cannot explain dark matter or dark energy, which together is said to make up 95 percent of the Universe. The new model of cosmology as postulated in this book holds that cosmic expansion gives birth to the Accelerating Expansion force of the Universe. This AE force creates inertial gravity in big things and quantum gravity in small things, thus giving us a *Final Theory Of Everything*.

Universe of big things and the micro Universe of small things. By bringing together UI gravity (Universal Inertial gravity) and quantum gravity under the umbrella of the AE force, we have a *Final Theory Of Everything* upon which to base our knowledge of cosmology and have a better understanding of the Universe.

The take home message: The standard model of cosmology is baseless because it depends on an ether and it cannot explain dark matter and dark energy, which together are said to make up 95 percent of the Universe. The new model of cosmology postulated in this book holds that cosmic expansion owes birth to the Accelerating Expansion force of the Universe or the AE force and its Inertial gravity in big things and quantum gravity in small things, thus giving us a *Final Theory Of Everything*.

What keeps galaxies together?

This section reveals how galaxies are kept together, a subject that has puzzled cosmologists for many years.

As bigger telescopes and more sensitive instruments peer deeper into space, new cosmic conundrums are making themselves known. Now that we have a *Final Theory Of Everything* it becomes possible to unravel some of these conundrums and mysteries of the Universe.

For decades cosmologists have been mystified by what keeps galaxies together. It is thought that about a billion years after the Big Bang, gravity caused atoms to gather into huge clouds of gas, forming collections of stars known as galaxies. So gravity is thought to be the force behind the formation of galaxies.

However, physicists and astrophysicists have long said that galaxies in our Universe rotate so fast that the gravity generated by their matter cannot theoretically hold them together and they should tear themselves apart, but they are still intact. This makes scientists speculate that there is a phenomenon, which we cannot see, that holds galaxies together. It's one of the great outstanding mysteries in cosmology and astrophysics: how exactly are galaxies held together?

Thus, even though the Universe has always been expanding, it is speculated that matter was pulled into lumps that eventually grew into galaxies like our own Milky Way. But what is holding galaxies together? Where is the glue?

Cosmologists study the birth of galaxies in an expanding Universe by running huge computer simulations, like the giant 'Millennium Run' performed by Durham University scientists.

The results of such computer simulations are then compared to the observed large-scale structure of the Universe.

It has become clear that as the Universe evolved, areas of greater density attracted more matter and grew bigger over time thus forming galaxies. This in turn created ever growing empty spaces between galaxies.

Physicists today say that it's unclear how the first massive galaxies could form so early after the Big Bang, and why some clusters of galaxies are connected by 'filamentary structures' and others are not (and how the filamentary structures were formed).

"In cosmology, filamentary structures are the largest known structures in the Universe, consisting of walls of gravitationally bound galaxy superclusters. These massive, thread-like formations can reach up to 260 million light-years across" (source: Galaxy Filament, Wikipedia.org).

The FTOE (*Final Theory Of Everything*) tells us that everything that exists has gravity, and hence has a gravity mantle. As soon as the Big Bang occurred, the AE force (Accelerating Expansion force of the Universe) sprang into action, giving gravity to everything that existed as explained in this book.

As stars and objects were formed in the early Universe their gravity mantles interconnected and they coalesced into the variety of galaxies and filamentary structures that we see today.

The aforementioned filamentary structures were formed in the same way as galaxies were formed, except that they are long chains of galaxies that form non-rotating structures. That is, the filamentary structures as a whole don't rotate, but within such structures, galaxies do rotate.

"Each of the galaxies in the filaments amounts to no more than a speck of dust on the grand scale, and they're not only rotating but moving along the tendrils as if they're pipelines. They move on helixes or corkscrew-like orbits, circling around the middle of the filament while travelling along it" (source: Noam Libeskind, Universe Today).

The key point here is that the interconnectivity of gravity mantles explains how all galaxies were formed, whether as rotating galaxies or as filament galaxies.

As mentioned, it is argued by some that galaxies rotate so fast that the gravity generated by their mass theoretically should not hold them together and they should tear themselves apart or fly off, breaking apart the galaxy. This is a false argument. If it were true, the formation of galaxies would not have been possible.

Cosmologists have been studying the 'galaxy rotation problem' for decades. Why do stars on the edge of galaxies stay put instead of flying off? They don't 'fly off' or tear themselves apart because there is no centrally dominating mass that keeps galaxies together through some kind of super powerful force of gravity. The false assumption that galaxy rotation is caused by the pull of a 'central mass' has created a 'galaxy rotation problem' where none exists. As already discussed, galaxies rotate because they are born from movement created by the AE force. This movement leads to angular momentum and rotation, and the formation of galactic disks.

Here is a quote from 'Universe Today' (Why does the Milky Way rotate? univesetoday.com):

"As cosmic clouds collapse, they form rotating disks. The rotating disks attract more gas and dust with gravity and form galactic disks. Inside the galactic disk, new stars formed.

What remained on the outskirts of the original cloud were globular clusters and the halo composed of gas, dust and dark matter. We know that galaxy rotation is happening because the Milky Way is a flattened disk, in the same way that the Solar System is a flattened disk. The centrifugal force from the rotation flattens out the galactic disk". **Note:** in the section that follows 'The Myth of Dark Matter', the existence of dark matter is questioned.

The centrifugal force of a galaxy is an 'outward' force (pushing things outward). It is created from the rotational movement of a galaxy, and this outward force is what gives rise to the mentioned 'galaxy rotation problem'. It is theorised that the combined gravity of a galaxy is not strong enough to counterbalance the outward centrifugal force. But this is false reasoning because there is no combined gravity of a galaxy. That is, there is no all-powerful gravity that is equivalent to the sum of all the gravities of objects inside a galaxy.

This false assumption (that the combined gravity of a galaxy is not strong enough to cancel out its centrifugal force) has led to the idea that dark matter must exist. That dark matter provides the extra mass needed to boost the combined gravity of a galaxy sufficiently to cancel out the outward centrifugal force, i.e. sufficiently to prevent a galaxy from flying apart.

But the 'glue' that holds galaxies together is not some immense force of gravity - such gravity has never been detected. If an incredibly powerful galaxy-wide force of gravity existed then it should be around us since we are part of the galaxy, but such omnipotent gravity is nowhere to be seen or felt.

Gravity mantle connectivity

Here is how galaxies stay together:

Every object in a galaxy will be under the influence of one or more gravity mantles at any one moment in time (for more on gravity mantles see the section 'Gravity Mantle'). This mantle connectivity is what holds galaxies (and clusters of galaxies) together. The interconnection of gravity mantles creates an enormous connectivity strength. Think of a single strand of cotton - it can easily be broken. But when several strands of cotton are intertwined they cannot be easily broken. This mantle connectivity is why the centrifugal force of a galaxy's rotation does not make stars go flying off or be torn apart.

But there is also another force that holds galaxies together. As explained in the section 'The AE force and gravity', cosmic expansion exerts a pushing force on everything that exists. So as cosmic expansion puts more space between galaxies it pushes galaxies apart from each other. This pushing force (the AE force) creates inertial gravity around objects. In the case of galaxies it means a galaxy as a whole will experience a pushing force from ***all sides to the galaxy***, akin to the inertial gravity bestowed on a star or a planet. This helps to keep stars in place on the periphery of galaxies, ensuring they do not succumb to centrifugal forces and go flying off. In other words, the same AE force that gives gravity to stars and planets also gives gravity to galaxies as a whole, acting as the glue that keeps galaxies together.

Note: When the AE force pushes all sides of a galaxy inward, thus helping to keep the galaxy together, this is known scientifically as a centripetal force, the opposite to a centrifugal force.

As stated previously, the AE force creates a continual pushing apart force between and around all objects, including subatomic particles and including whole galaxies. The pushing apart force can be overcome by the pushing together effect of UI gravity if they are close enough to be affected by each other's gravity. This is what holds atoms together in the human body, or stars together in a galaxy.

This battle between pushing apart and pushing together forces is widely accepted in contemporary cosmology and it goes like this:

"Hubble's law states that if two objects that are stationary to each other had no force between them, and were left alone, the distance between them would increase with time because space itself is expanding. This is what Hubble's law addresses. In the case of the Milky Way and Andromeda galaxies (and all galaxies for that matter) there is a force between them: gravity. The gravitational force between the Milky Way and Andromeda galaxies has produced an acceleration that is causing the two galaxies to be moving towards each other faster than the space between them is expanding as calculated by Hubble's law. However, the vast majority of galaxies lie far enough away from the Milky Way that the gravitational force between us and them is small compared to the Hubble expansion, and as a result Hubble's law dominates. In short, Hubble's law applies throughout the Universe, but localised systems may have enough gravitational attraction between them that the gravitational effects dominate".

Coming back to gravity mantles, the following image gives an idea of how gravity mantles overlap, but of course, everything is 'fluid' and in constant movement. So the gravity-mantle-overlaps are constantly moving around and changing. The size of gravity mantles varies enormously, depending on the mass of the object inside each gravity mantle. And just about every object in a galaxy will be caught simultaneously by several gravity mantles as depicted in the following image:

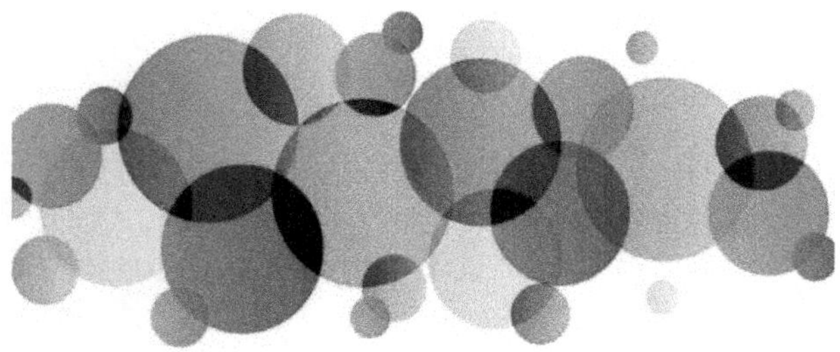

Another way to think of these gravity mantles is to imagine a round rug. The threads of the rug represent the overlapping gravity mantles which act as 'ties'. The overlapping gravity mantles tie the galaxy together as a whole:

Planets orbiting a sun, and moons orbiting a planet, will be totally immersed in the gravity mantle of whatever they are orbiting as otherwise the object in orbit would go 'flying off' and stop orbiting. For example, the Earth is totally located inside the Sun's gravity mantle, and our Moon is totally immersed inside the Earth's gravity mantle. But the Sun's gravity mantle also overlaps the gravity mantle of Proxima Centauri (our nearest star).

Equally, a whole galaxy will have its own overall gravity mantle because of everything staying together inside the galaxy. As such, the gravity mantle of galaxy 'A' can overlap the gravity mantle of galaxy 'B'. Clearly, inside any galaxy, all gravity mantles within a galaxy will be constantly moving around and weaving in and out of other gravity mantles around them because everything is always moving.

It is estimated that our galaxy has a mass of more than 1 trillion suns, with a force of gravity that extends out more than 600,000 light-years; a third of the distance to the nearby Andromeda galaxy. But the gravity mantles of the Milky Way and Andromeda will be large enough to overlap thus causing both galaxies to gradually coalesce (come together).

So now we know why objects in galaxies stay together, and why clusters of galaxies stay together. There is no incredibly strong gravity that comes about from the sum of all the small gravities inside a galaxy, or from the sum of all the masses inside a galaxy.

We also know why stars on the periphery of rotating galaxies are not torn apart or thrown out; it's because such stars are firmly 'tethered' to the galaxy-wide interconnected gravity mantle. In other words, the gravity mantle of a galaxy as a whole ensures that objects within a galaxy stay put. This is reinforced by the AE force that exerts a pushing force to all 'sides' of a galaxy, thus helping to keep galaxies together.

And last but not least another mystery is unravelled. It is thought that the Universe is nearly 13.8 billion years old, yet galaxies were formed in the first billion years of the Universe. Cosmologists are puzzled - how were they formed so quickly?

The answer is that soon after the Big Bang, everything that existed at the time was endowed with a gravity mantle due to the AE force. As mentioned, the radius of gravity mantles can be very extensive, thus readily making gravity mantles overlap and connect. Such mantle connectivity in turn made it possible for stars, planets and galaxies to be formed soon after the Big Bang. And this formation of galaxies is still going on today.

The Myth of Dark Matter

Here is an abridged extract from 'Dark Matter', Wikipedia.org:

Quote
"Dark matter is a hypothetical form of matter thought to account for approximately 85% of the matter in the Universe. Many so-called 'experts' think that dark matter is abundant in the Universe and has had a strong influence on its structure and evolution.

The primary evidence for dark matter is said to come from calculations showing that many galaxies would not stay together if they did not contain a large amount of unseen Dark Matter. This mysterious Dark Matter is thought to be composed of as-yet-undiscovered subatomic particles - some new kind of elementary particle. Many experiments to detect and study dark matter particles directly are being actively undertaken, but none have yet succeeded".
Unquote

The reasoning put forward by those investigating Dark Matter is that the velocities of stars orbiting around galaxies seem too high. At such speeds they should fly off or be torn apart, thus preventing the steady formation of galaxies.

To explain why this does not happen it is theorised that there is extra matter (i.e. Dark Matter) that acts to hold things together inside a galaxy. And that this process can be repeated on the scale of clusters and superclusters of galaxies. So it is thought that there is not enough visible matter to keep things together inside a galaxy and that there must be some sort of dark matter that gives extra mass to a galaxy so as to give it greater gravity.

But how exactly does dark matter exert a pulling force of gravity so as to keep galaxies together? There is no explanation for this. The nearest you get is this: dark matter adds to the combined mass of a galaxy, and greater mass exerts greater gravity.

Without knowing why, we know from measurements that the gravitational force is directly proportional to the mass of both interacting objects; more massive objects will attract each other with a greater gravitational force. So as the mass of either object increases, the force of gravitational attraction between them also increases. But why?

According to theory, the reason mass is proportional to gravity is because everything with mass emits tiny particles called gravitons. These gravitons are responsible for gravitational attraction. The more mass, the more gravitons. The problem is that gravitons have never been detected, and no research has been able to show that they exist.

And what about Newton's Law of Universal Gravitation you might well ask? Isaac Newton famously decreed that the nature of gravity is universal (is the same) throughout the Universe. He also said that the force of gravity is directly proportional to the mass of both interacting objects.

For example, if the mass of one of the objects is doubled, then the force of gravity between them is doubled. If the mass of one of the objects is tripled, then the force of gravity between them is tripled. If the mass of both of the objects is doubled, then the force of gravity between them is quadrupled; and so on.

Similarly, Newton said that if the separation distance between two objects is doubled (increased by a factor of 2), then the force of gravitational attraction is decreased by a factor of 4 (2 raised to the second power), and so on. But Newton was never able to explain why this was so. Exactly why does mass cause gravity between objects?

Newton's theory of gravity is regarded as being the most accurate in modern-day physics. It is not 100% accurate because Newton's Law of Gravitation does not take into account kinetic energy because it was unknown at the time. At very high speeds kinetic energy adds to the amount of mass of the object that is moving. This means that Newton's Law of Gravitation can only give approximations of the force of gravity at very high speeds such as the speed of light or the speed of subatomic particles.

When scientists add the effect of kinetic energy to Newton's Law of Gravitation they are able to obtain very accurate calculations (99.999%) for the force of gravity.

In short, today's science cannot explain at a fundamental level exactly why greater mass equals greater gravity (even though it can be calculated), so the scientific consensus resorts to relativity, saying that greater mass acts to make a greater curvature of space, and hence greater gravity.

But this begs the question: How can the combined mass of a galaxy, including dark matter, be kept together within a curvature of space? Exactly what parts of a galaxy's space is being curved so as to keep the whole galaxy together? Answers to these types of questions are not forthcoming.

Note: As explained in the section 'Space Navigation' using relativity to calculate the force of gravity can only ever give rough approximations and that is why Newton's Law of Gravitation (rather than relativity) is used in space navigation.

Now, for the first time, this book reveals at a fundamental level exactly why greater mass equals greater gravity. Here is the answer:

The FTOE tells us that cosmic expansion creates movement, causing objects to either separate or coalesce. This AE force of movement creates inertial gravity in big things and quantum gravity in small things.

So at a fundamental level the nature of inertial gravity arises from the force of inertia (also called 'inertial force') caused by movement emanating from cosmic expansion. Similarly, at a fundamental level the nature of quantum gravity arises from the force of electromagnetism caused by movement emanating from cosmic expansion. Quite simply, any kind of movement creates inertia and hence gravity. And since everything in the Universe is always moving, it means everything is subject to gravity.

The nature of the aforementioned UI gravity (Universal Inertial gravity) is supported by the mathematics - see the section titled 'The Mathematics of the FTOE' in this book. Now that we know the fundamental nature of quantum gravity, existing mathematical tools can be used to better understand and calculate quantum gravity, and reconcile quantum gravity with UI gravity.

Dark matter is not needed to explain how galaxies are kept together. Let's not forget the growing research showing that dark matter is a myth: *"A simple test suggests that dark matter does not in fact exist. If it did, we would expect lighter galaxies orbiting heavier ones to be slowed down by dark matter particles, but we detect no such slow-down. A host of other observational tests support the conclusion: dark matter is not there"* (source: Pavel Kroupa, head of the Stellar Populations and Dynamics research group, professor of astrophysics at University of Bonn, Germany, July 2022).

Another main argument for the existence of dark matter relates to the bending of light. It is speculated that as light travels through a region of dark matter, its path gets distorted by gravity. Instead of taking a straight line, the light is bent back and forth depending on how much dark matter the light passes through. However, light can never bend or fall prey to gravity emanating from objects it may go past.

If dark matter were to be making light change direction, thus making it appear as if it is bending, it implies two things. (1) That dark matter is made of particles that have mass, and it is this mass that has a gravitational effect on light (no such particles have ever been detected).

And (2) if light is made to bend from regions of space with dark matter, it must mean that such light is travelling through parts of space that are completely devoid of gravity (except for dark matter particles). How can it be known that such parts of space are completely devoid of gravity? As explained in the sections 'Gravity Mantle' and 'What keeps galaxies together' the effect of gravity mantles across space can be enormous, overlapping between galaxies.

Quite simply, we don't need dark matter to explain how galaxies are kept together. We now know that galaxies are kept together through the interconnectivity of gravity mantles, as discussed in the previous section, and through the AE pushing force.

The take home message: Galaxies are held together by UI gravity that acts as a pushing force to all sides of a galaxy, and gravity mantles serve to keep things together inside a galaxy. Dark matter is debunked and has no role in keeping galaxies together.

Unravelling Relativity

This section discusses various aspects of special and general relativity, showing them to be spurious.

Einstein's general theory of relativity is all about gravity and how it works. Put briefly, the theory says that time and space are fused together in a phenomenon known as spacetime. Within this theory, objects cause spacetime to curve, and gravity is simply the curvature of spacetime. The theory requires you to believe or accept that when an object is close enough to another object the space between them becomes 'curved'. It is this curvature of space that causes the two objects to move towards each other.

The degree of curvature is determined by the amount of mass of the objects. So the mass of an object itself is what distorts and curves space. But how?

To date, no explanation has been forthcoming that shows the precise mechanism by which the mass of an object can curve space around itself in all directions. According to Einstein's theory of gravity, every object in the Universe makes the space around itself curve to a greater or lesser extent. Does this mean that an isolated object in space that is not near anything else has curved space around it? Or does curved space only make an appearance when two or more objects are close enough to each other? If so, how close? And how can space – emptiness, nothingness – have any properties at all? And if galaxies are kept together by gravity (with or without dark matter) does this mean an overall curvature of space is applied to a whole galaxy, and that this overall curvature of space contains trillions of smaller space curvatures endowed to every object inside a galaxy?

None of these questions have been fully explained by science. Einstein himself was unable to answer these questions at a fundamental level. He pondered long and hard about the existence of an ether (an all-pervading presence of invincible subatomic particles everywhere in space).

Einstein needed to explain how empty space between, say, two objects become curved when they are close enough to each other but not when they are far from each other. How can the two objects (with nothing between them) detect each other's presence so as to trigger space curvature?

To resolve this conundrum Einstein eventually and reluctantly had to state that an ether exists; that as objects approach each other they disturb the ether thus emitting 'gravitational waves' and this triggers a curvature of space. But to date no ether has been detected anywhere in the Universe. And without an ether the general theory of relativity completely falls apart.

Those who profess the validity of relativity in today's world get round this conundrum by referring to dark matter. This mysterious substance that has never been detected is the modern-day equivalent of an ether. Dark matter is said to make General Relativity work by causing a disturbance in space between objects such that a curvature of space is triggered. Equally, dark matter is needed to explain so-called gravitational lensing.

If dark matter exists it should be everywhere, if not why not? If it's everywhere, why can't we detect dark matter in laboratories on Earth? If the particles of dark matter are too small to be detected by current science, then how can they be responsible for the bulk of mass in the Universe?

The following topics are discussed in what follows:

Gravitational waves.
Gravitational tides.

Perihelion precession of Mercury.
Space navigation.
Testing General Relativity.
Testing Special Relativity.
Spaceship Thought Experiment.
Types of light speed (summary).
Is special relativity used in daily life?

Gravitational waves

Einstein predicted gravitational waves in his general relativity, describing such waves as 'ripples in spacetime'. Gravitational waves are meant to occur when objects are close enough to curve space between them, and in doing so an all-pervading ether or dark matter is disturbed so as to produce gravitational waves.

"Despite high expectations to discover gravitational waves, the LIGO and VIRGO detectors have been producing nothing more than false alarms. With the award of the 2017 Nobel Prize in Physics to Rainer Weiss, Barry C. Barish, and Kip S. Thorne, the question of whether gravitational waves have been directly detected seemed to be settled forever. However, all attempts to detect gravitational waves have failed to date" (source: Gravitational Waves: The Silent Disaster, Real Physics, 2020).

The failed prediction of gravitational waves by General Relativity is yet another nail in the coffin of spacetime and the curvature of space.

Gravitational tides

Tides are caused by the gravitational effect of the moon and the sun. The rise and fall of the tides have played a crucial role in the development of the human species, in weather patterns, in the natural world and in many maritime-related activities.

Unfortunately, the following two images show how classical tidal theory is currently taught in classrooms, but such teachings are wrong on several counts.

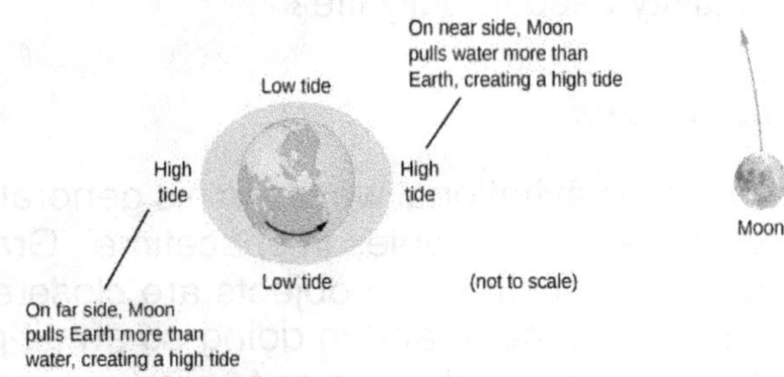
On near side, Moon
pulls water more than
Earth, creating a high tide
Low tide
High
tide
High
tide
Moon
Low tide
(not to scale)
On far side, Moon
pulls Earth more than
water, creating a high tide

Here is another erroneous image attempting to explain tide theory:

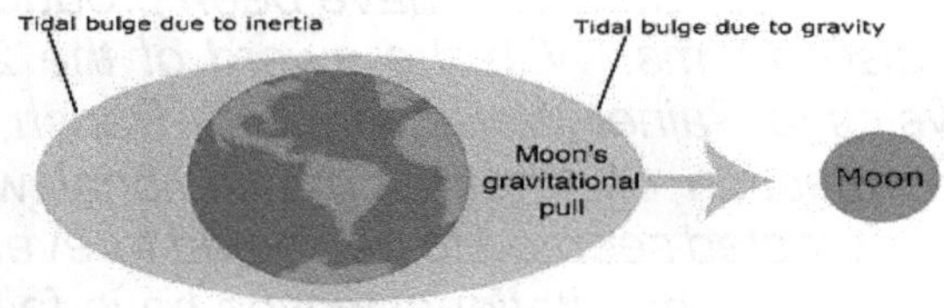
Tidal bulge due to inertia
Tidal bulge due to gravity
Moon's
gravitational
pull
Moon

We can see in the above two images the following incorrect ways that classical wave theory is currently taught:

1. The moon does not pull Earth's oceans anywhere. Gravity is not a pulling force because this implies that some mysterious force reaches out from the moon towards earth to 'grab' and pull the oceans. A pulling force of gravity does not exist. The moon does not pull the oceans from the nearside, nor does it pull the Earth from the far side.

2. There is no 'line of gravity' that goes from the moon across the equatorial regions of Earth so as to exert a greater force

of gravity along Earth's equator (so as to create ocean bulges at each side of the equator).

3. Earth is not being pulled towards the moon thus leaving behind a bulge of water on the far side due to inertia.

"When teachers explain ocean tides, they frequently describe how the Moon's gravity pulls on Earth and all of its water. This, they often say, leads to a gravitational imbalance, which stretches the ocean into two opposing bulges: one that's closest to the Moon (where the Moon's gravity is strongest); and one on the opposite side from the Moon (where its gravity is weakest). The bulges do occur, but this kind of explanation for the bulges is wrong. Our tidal bulges are actually the product of a complex dance of gravity between the Moon, Earth, and Sun. And the total effect is more of a 'push' than a 'pull' on Earth's water" (source: Gabe Perez-Giz, astronomer and astrophysicist at NYU, The Moon's Gravity Does Not Fully Explain How Ocean Tides Work, PBS Space Time).

Ocean water covers about 71 percent of Earth's surface and is connected as one liquid body. When gravity from the Sun and Moon line up in a certain way vis-a-vis Earth, the liquid body of earth is squeezed, resulting in the bulges to either side of the planet.

Ocean bulges do not appear at the north and south poles because of a combination of three factors: (1) Earth's gravity is stronger at the north and south poles, (2) Earth's centrifugal force is weaker at the poles, and (3) the Arctic is surrounded by land and is almost landlocked, while Antarctica is land surrounded by ocean on all sides. For these reasons there is no noticeable bulge at the poles. It is not known whether the frozen nature of the poles may also help to prevent bulges in these regions.

The *Final Theory of Everything* postulated in this book fits in nicely with tidal wave theory because UI gravity (Universal Inertial gravity) is a pushing force. As explained in the section 'The Coalescing Force', inertial gravity is created around everything in the Universe. When the gravity mantles of the Moon and Sun 'encase' the whole earth at certain times, UI gravity exerts a pushing force to all sides of Earth, albeit the tidal influence of the Sun is less than that of the Moon.

This in turn pushes and squeezes the oceans to bulge out to both sides of the planet. As mentioned, there is no bulge to the north and south poles because of Earth's geology.

In effect Earth is being squeezed from all sides, and the oceans being a fluid, bulge out to both sides of the planet as depicted in the following image:

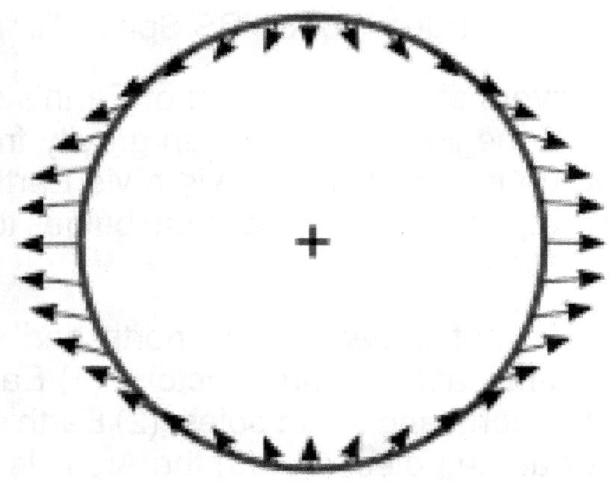

"The overall effect of these tidal forces is to 'squeeze' the oceans, and produce two tidal bulges on opposite sides of the Earth—one facing the Moon and a slightly smaller one facing away from the Moon" (source: Vigdis Hocken, et al, What Causes Tides? timeanddate.com).

These two bulges combined with Earth's rotation cause the tides that we see on Earth. **Note:** tides also occur in smaller masses of water such as lakes, but the tidal effects are too small to be noticeable.

What about relativity you might well ask? How does general relativity account for the tides?

When Einstein published his general theory of relativity he thought long and hard about the gravity of tides and how to reconcile this with his general relativity. The problem he faced was that the curvature of space is neither a pushing or a pulling force - it is a curved 'pathway' that when followed brings objects together because of the curvature, as depicted in the following image:

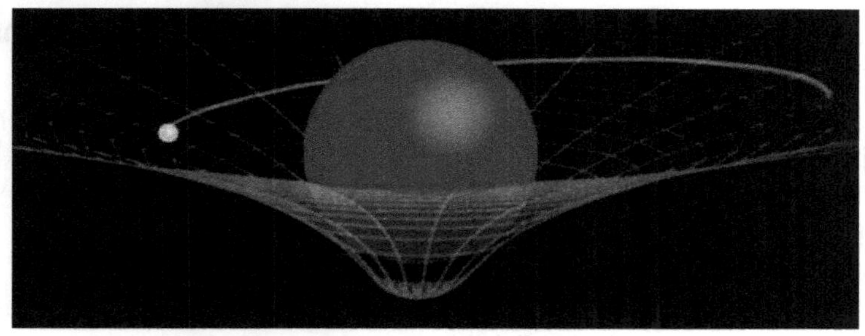

According to general relativity, when a smaller mass passes near a larger mass, it curves toward the larger mass because spacetime itself is curved toward the larger mass. The smaller mass is not 'attracted' to the larger mass by any force. The smaller mass simply follows the structure of curved spacetime near the larger mass.

The big problem for Einstein was that he had no explanation as to how the oceans bulged out to both sides of the planet under the influence of the Moon and Sun. If spacetime curvature is merely a direction of travel (without any kind of pushing or pulling force whatsoever) then how can the bulge of the oceans (and hence tides) be accounted for?

Enter 'test particles'. If you google 'test particles' you will discover that a test particle is a theoretical particle, an idealised model of an object whose physical properties (usually mass, charge, or size) are assumed to be negligible except for the property being studied, which is considered to be insufficient to alter the behaviour of the rest of the system.

Put simply, a test particle is an imaginary particle with no mass or size, and is used to 'test' physical phenomena. The concept is that a test particle often simplifies the understanding of problems, and can provide a good approximation for physical phenomena. The physical existence of test particles is not proclaimed and they have never been seen or verified.

To make general relativity 'work' with tides, Einstein introduced test particles into his field equations. Here is a description of how tides can be accounted for in general relativity (abridged compilation from several sources):

Quote

Tidal forces in general relativity arise because in general relativity test particles which are not acted on by any other forces at all, move along geodesics in spacetime. The presence of mass, such as the moon and the earth, acts to curve spacetime and the test particles then move along curved paths in space. Because the curvature is not uniform in the spacetime around the earth and moon, test particles which start out very close to each other eventually begin to fall along slightly different paths - the geodesics they follow in spacetime begin to deviate from each other.

This is described by an equation called, unsurprisingly, the equation of geodesic deviation. The deviation of particles moving along nearby geodesics in spacetime appears to us exactly as if a force is acting on the particles, providing we just look at the motion for a very short time and over a short distance.

Unquote

With this, general relativity was said to account for tides. To accept the above explanation for how general relativity accounts for tides you have to accept the existence or at least the effect of so-called 'test particles'. But these imaginary particles have never been shown to exist in any real sense.

More significantly, if spacetime curvature accounts for tides, then how does spacetime curvature make the planet's oceans bulge, bearing in mind that spacetime gravity is not a force, it exerts no kind of pulling or pushing force at all, it is merely a curved pathway?

Let us assume for a moment that when the Moon and the Sun are aligned on Earth so as to produce ocean bulges, this triggers a spacetime curvature of space between all three bodies. In this scenario, the moon will be following a curved path that takes it closer to Earth and then continue on its elliptical orbit. And in this scenario, since spacetime curvature exerts no kind of pushing or pulling force of gravity, one has to wonder how spacetime curvature can account for the oceans to bulge to each side of Earth?

Perihelion precession of Mercury

The phrase 'perihelion precession of Mercury' roughly translates to:

The effect of the Sun's gravity in shifting Mercury's orbits.

Mercury has a very pronounced elliptical orbit around the sun due to various gravitational effects from the Sun and nearby planets. Each time Mercury's orbit takes it close to the Sun, the planet receives a 'gravity kick' that sends Mercury into a new (shifted) elliptical orbit.

Einstein reasoned that as Mercury moves toward the Sun, it moves deeper into the Sun's space-time gravity well, and this in turn makes Mercury follow a new (shifted) elliptical orbit.

"Einstein showed that general relativity agrees closely with the observed amount of perihelion shift. This was a powerful factor motivating the adoption of general relativity" (source: Tests of general relativity, Wikipedia.org).

Einstein proposed that the perihelion precession of Mercury be measured using the field equations presented in his 1916 paper so as to test his general theory of relativity. This was done several times vis-a-vis Mercury's orbits showing that his equations were accurate. The following seven points arise:

1. Relativity is not required to explain the perihelion precession of Mercury. It is easily explained by the effects of gravity from the Sun and nearby planets. And furthermore all orbiting planets in the Universe undergo a perihelion precession; it is just that Mercury's perihelion precession (i.e. shifting elliptical orbit) is very pronounced.

2. Today's astronomers can measure the perihelion precession of Mercury very accurately using modern astronomical technology that does not involve any kind of Einsteinian relativity. And furthermore, General Relativity can only ever give approximations of perihelion processions. The mentioned 1916 paper was a set of manipulated equations specifically designed to give an exact result for the perihelion precession of Mercury, and Mercury only.

This is so because the field equations of relativity are non-linear, they can never be completely solved, they can only ever give approximations for calculating orbital arcs (an arc is the amount of shift from one elliptical orbit to another).

3. Normally, the amount of orbital ark from one orbit to another is nearly the same. But in the case of Mercury each orbital arc (each shift into a new orbit) is very pronounced. This discrepancy was calculated to be 43 seconds of arc per century, and this was well-known before Einstein's time. Why so? Because Newton's equations, taking into account all the effects from the other planets (as well as a very slight deformation of the sun due to its rotation) and the fact that the Earth is not an inertial frame of reference, predicts a precession (change) of 5557 seconds of arc per century. This amounts to a *discrepancy* of 43 seconds of arc per century. But Newton's equations could not explain mathematically the cause of the 43" arc anomaly.

4. Einstein knew about Newton's 43" arc anomaly and indeed based his own arc calculations on Newton's work. Einstein wanted to present a method for calculating Mercury's 43" arc anomaly (but based on his relativity field equations) so as to prove the veracity of his general theory of relativity. But given the nature of relativity, Einstein's field equations could only give an approximation for the perihelion precession of Mercury (not an exact calculation showing how the 43" arc anomaly arises).

5. To resolve Einstein's conundrum he collaborated with Karl Schwarzschild, a German physicist and astronomer of the time. Schwarzschild reviewed Einstein's equations and he came up with a solution to make Einstein's field equations produce an exact result for the 43" arc anomaly rather than just an approximation. Schwarzschild did this by employing a mathematical 'trick' to make Einstein's Field Equations

provide a closed (exact) solution relating to the perihelion precession of Mercury.

"Schwarzschild found a simple trick that allowed him to avoid the problem of the non-allowed coordinates [of spacetime]. *He came up with new variables for the spherical coordinates of the determinant 1 to the power of 21. Einstein's field equations and the coordinate condition of the square root of minus g from his November 18 paper were satisfied. So returning back to the non-allowed spherical coordinates, we arrive at the exact solution to Einstein's problem, and to the mathematical singularity in the solution when R = 0".* Source: Galina Weinstein, 'Einstein-Schwarzschild-the Perihelion Motion of Mercury and the Rotating Disk Story', Tel Aviv University, 2014.

Note: The 'Schwarzschild Metric' as it is known today can accurately measure the amount of shift in planetary elliptical orbits. But the Schwarzschild Metric is a set of equations based on (but ***not the same as***) Einstein's General Relativity field equations. Thus, the Schwarzschild Metric shows that relativity field equations can only ever give rough approximations for measuring the movements of astronomical objects unless the field equations are adapted (changed) to the Schwarzschild Metric.

6. In short, Einstein manipulated his relativistic equations (with help from Schwarzschild) so as to make such equations give an exact result for a particular instance, i.e. an exact result that corresponded with Mercury's orbital anomaly. Furthermore, in manipulating the said equations Einstein used covariance algebra, a type of mathematics that is mostly discredited today. It is discredited mainly because covariance is not a good way to measure the strength of a linear relationship because it is not invariant to deterministic linear transformations. As such, covariance allows a wide range of

results from which you can choose the result that most suits you.

7. In 1915 Einstein published his general theory of relativity that we are all familiar with. But it can only give approximations for calculating Mercury's 43" arc anomaly. But back in 1915 Einstein wanted to show that his general theory involving spacetime curvature could give an *exact* calculation of the 43" arc anomaly. To clarify, he was not predicting the occurrence of the 43" arc anomaly (a popular misconception) as the anomaly was already well-known, rather he was showing that relativity can be used to calculate the 43" anomaly precisely.

So as his 1915 paper on General Theory could only give approximations for the 43" anomaly, he needed a mathematically exact solution specific to mercury. As mentioned, he collaborated with Schwarzschild and in 1916 Einstein presented a version of his general relativity, but with the mathematics manipulated specifically for the perihelion precession of Mercury. At the time this turned out to be a powerful factor in motivating the adoption of general relativity in the absence of some other credible theory of gravity.

Thus when Einstein proposed that his general theory of relativity could be used to predict exactly the amount of the perihelion precession seen in Mercury, he already knew well beforehand that he would get the precise result he was predicting having rehearsed his contrived calculations. In contemporary parlance it was a 'setup'.

"Einstein's general theory of relativity has yet to be tested in a regime that will set it apart from Newtonian Gravity" (source: Nasa Education Program, Newton Einstein FactCards.pdf).

Space navigation

Does NASA use Newtonian physics? The answer is YES because it is perfectly accurate for space navigation, and in fact for any measurements in the cosmos, except for things going at or nearly the speed of light. This begs the question *why isn't General Relativity (GR) used in space navigation?*

The consensus answer by those defending Einsteinian relativity is that Newtonian physics is sufficiently accurate so there is no need to use general relativity physics, which would be overkill. This is not so. Newtonian physics is used because it is 'deterministic', i.e. you can determine the output (the result) from a given starting condition.

A deterministic approach is absolutely essential in space navigation because you want to control and know where the navigation will go (what will happen) in the most accurate way possible.

In GR (General Relativity) you cannot do that because general relativity is not deterministic. Compared to Newtonian physics, with *"General Relativity, the challenge is much greater. Even if you knew those same pieces of information — positions, masses, and momenta of objects — plus the particular relativistic reference frame in which they were valid, that wouldn't be enough to determine how things evolve. The structure of Einstein's greatest theory is too complex even for that"* (source: Ethan Siegel, This Is Why Scientists Will Never Exactly Solve General Relativity, Forbes.com, 2019).

Here's another way to look at it. In considering GR it isn't the net force acting on an object that determines how it moves and accelerates, but rather the curvature of space (and spacetime) itself. This immediately poses a problem, because the entity that determines the curvature of space is all of the matter and energy present within the Universe, which includes a lot more than merely the positions and momenta of the objects that we need to negotiate in space navigation.

General Relativity poses a unique set of challenges that don't arise in a Newtonian Universe. Such challenges include:

* A curvature of space that is continuously changing.

* Every mass has its own self-energy that also changes spacetime's curvature.

* Objects moving through curved space interact with it and emit gravitational radiation.

* All the gravitational signals generated only move at the speed of light, so that the object's velocity relative to any other object results in a relativistic (length contraction and time dilation) transformation that must be accounted for.

When you take all of these factors into account, it all adds up to equations that are too complex and unreliable even for powerful computers because we humans cannot give computers the reliable input required for such calculations (garbage in, garbage out).

GR can only ever give approximate calculations for space navigation and is therefore much too unreliable. In short Newtonian physics is used because it is an accurate and practical solution for space navigation. General Relativity is much too ambiguous, inaccurate and impractical for space navigation.

Newton's law of universal gravitation states that every object attracts every other object in the Universe with a force that is proportional to the product of their masses and inversely proportional to the square of the distance between their centres.

Newtonian physics is not perfect but it suffices for space navigation. These are its main shortcomings:

- It doesn't explain the fundamental underlying nature of gravity. The FTOE (Final Theory Of Everything) fully explains this.
- It doesn't provide accurate measurements for major changes in planetary orbits (precession of the perihelion). Modern astronomy can easily make such measurements without relativity.
- It doesn't account for gravitational lensing. The FTOE fully explains this (see 'Comic lensing').
- It doesn't explain the role of gravity in keeping galaxies together. The FTOE fully explains this.
- It postulates that the effect of gravity of an object extends into space indefinitely. The FTOE denies this.

These Newtonian shortcomings do not at all affect space and optical-space measurements using Newtonian physics. It is only when you approach the speed of light that Newtonian physics is not so accurate. Why so? Because Newton did not take into account in his equations that the higher the speed of an object the greater its kinetic mass.

So unlike general relativity, Newtonian physics continues to be used today in all types of space navigation with full confidence.

Relativists claim that Newtonian physics is inaccurate for measuring the very high speeds of objects, and that Special Relativity is required for the accurate measurement of such speeds. The reasoning is that at very high speeds you need to take into account the slowing down of time and the contraction of length (of the object travelling) so as to obtain an accurate measurement of the speed of said object. This conflicts with Newton's laws of physics that remain the same at whatever speed you are travelling.

Such Special Relativity reasoning is disingenuous to say the least. The physics of today can easily adapt Newtonian equations to take into account high speed kinetic energy. And furthermore, as mentioned, relativity can only ever give approximations when measuring any kind of speed.

Testing General Relativity

At various points in this book, and in the context of postulating the *Final Theory Of Everything*, Einsteinian relativity has been discussed. In physics both special and general relativity have been questioned in various ways ever since such theories were first put forward by Albert Einstein many years ago.

Whole books and many articles have been published on the subject of testing and challenging relativity. From this, the picture that emerges is that most of the claims put forward by relativity remain unproven or they are shown to be incorrect. We won't attempt to discuss this subject in detail as it would be much too involved and irrelevant in the context of the theme of this book, namely a *Final Theory Of Everything.*

To summarise, GR (General Relativity) falls apart and is shown to be spurious for the following main reasons:

1. GR is entirely dependent on the existence of an ether for curved space to work. There is no evidence that an ether or that dark matter exists. Relativity says that without an ether, space curvature cannot be triggered, thus showing that Einsteinian gravity is spurious.

2. GR is entirely dependent on the existence of time as an absolute phenomenon (a separate agent in the form of a

‘fourth dimension’). However, this has never been verified or detected. Without the existence of time as an absolute independent phenomenon, the notion of gravity as postulated in GR falls apart.

3. No theory of gravity can work for just some parts of the Universe, and not for other parts. GR fails to explain black holes, it doesn’t work for subatomic particles, and it cannot account for maritime tides on Earth. Experts have long suspected that general relativity can't be right in the realms of extremely high mass density. At the centre of a black hole, for instance, the theory's equations no longer make sense, because they imply that the density of matter becomes infinite.

4. General relativity does not respect local energy-momentum, i.e. the energy and momentum of an object at rest. ‘At rest’ means being stationary relative to a particular frame of reference or another object. There are serious problems with local energy-momentum conservation in general relativity. It is well-known that Einstein's theory does not assign a definite stress-energy tensor to the gravitational field. This property is extremely unsatisfactory, because one knows that all other fundamental interactions in nature actually do respect the principle of local conservation of energy-momentum. Essentially, the non-existence of a stress-energy tensor is a consequence of the purely geometrical interpretation of gravity as a curvature of space-time.

Testing Special Relativity

Regarding SR (Special Relativity), it is a theory that explains how speed affects mass, time and space. It stipulates among other things that as an object approaches the speed of light, the object's mass becomes infinite and so does the energy required to move it. That means it is impossible for any matter to go faster than the speed of light. You are also asked to believe that the faster an object moves, the more it contracts in physical length.

If one attempts to analyse the veracity of SR you fall into a quagmire of convoluted confusion and contradictions. For example, SR forbids the existence of an ether, i.e. an ether of subatomic particles (with mass) moving around in space. Yet such an ether is an absolute requisite for general relativity, a clear conflict between special and general relativity.

At various points in this book examples are given of the relative speed of light. For example, if you sit in the back of a car going at 100 kph and you shine a torch forward we have the following scenario. Relative to the road, the torch light is going at the speed of light plus the speed of the car. As the car passes an observer at the roadside, the observer will see the torchlight go past at speed c + 100 kph. **Note:** the speed of light inside the moving car has not changed, it remains at speed c.

So from the observer's point of view the torchlight in the passing car is moving at a relative speed of c+100 kph. But SR denies this. Special relativity says that the torchlight goes past the roadside observer at speed c (not speed c + 100 kph).

A point of clarification: SR says that *“the speed of light in a vacuum is constant for all observers, regardless of their motion relative to the source”*. This causes endless confusion and misunderstandings. Obviously, the constant speed of light does not change by virtue of being observed. You, the observer, can be moving anyway you like, right next to the light or a million miles away - it will not change the constant speed of light. What special relativity is actually saying is that the ***relative*** speed of light is the same as the actual speed of light.

The relative speed of light is simply how fast the light is moving in relation to you or some other frame of reference. The actual (true) speed of light remains unchanged for all frames of reference. But for example if you were to measure the speed of light at c while you go in the opposite direction at speed x, then the speed of light *in relation to you* is c minus x (i.e. the *relative* speed of light in that particular instance is c minus x).

So when it is said that the speed of light in a vacuum is the same for all observers, this is correct. But it doesn't follow that the *relative* speed of light is the same for all observers regardless of their motion relative to such light. The denial of the relative speed of light in SR serves to show it is baseless and spurious.

Types of light speed (summary)

1. Direct light speed. Maxwell's constant speed of a light ray at c in a vacuum (300 million metres per second). This concept is a law of nature and stands true today. Direct light speed refers to the speed of the photons themselves that make up light.

2. Relative light speed. The constant speed of a light ray at c in a vacuum, which is also travelling relative to many bodies moving at different speeds. This concept describes the relative speeds of any light ray. The relative speed is the sum of two or more speeds or the difference of two or more speeds. It is a mathematical expression of the greater or lesser distance between light c and a given moving object. This concept stands true today.

3. Incident light speed. This is the overall journey-time of a given stream of photons that has been attenuated, i.e. that has been absorbed and then emitted as incident light. As a result of attenuation, greater time-intervals are put between moving photons which creates a longer journey-time for a whole stream of photons

4. Einsteinian light speed. This refers to Einstein's physically impossible constant speed of a light ray at c relative to all uniformly moving bodies in the Universe at the same instant, regardless of such bodies' different velocities relative to such light ray. This clever result of mathematical manipulation was Einstein's artificial and meaningless so-called solution for the paradoxes which he created.

Note: Maxwell, rather than Einstein, discovered the correct nature of light. In Maxwell's monumental work, physicist Richard Feynman made this comment regarding J. C. Maxwell in the context of light speed:

"From a long view of history there can be little doubt that the most significant event of the 19th century will be judged as Maxwell's discovery of the laws of electrodynamics. The American Civil War will pale into provincial insignificance in comparison with this important scientific event of the same decade." This comment surprised many at the time because Einstein was then extremely popular.

Einstein devised his special theory of relativity by using clever mathematical manipulations involving covariance algebra. Covariance refers to metrics that calculate the variance of two variables. As mentioned earlier, it is a kind of algebra that is mostly discredited because it is not a good way to measure the strength of a linear relationship, i.e. it is not invariant so as to determine linear transformations.

Note: Covariance became the main focus of Einstein's concepts and mathematics in both his Special and General Relativity. It gave him an amazing mathematical trick for manipulating his equations in his world of pseudo-scientific fantasies. But such magical tricks were nothing more than mathematical illusions. They have no physical meaning for the real world in which we live.

In order that his artificially constant velocity of any light ray at c (with respect to any moving body in the Universe) could remain mathematically consistent with all of the other concepts of physics, Einstein was also forced to mathematically distort Newton's laws of mechanics as well as most other laws of physics.

Einstein also tried to find other theories and experiments that might appear to confirm his distorted concepts and his unnecessary and invalid Special Relativity. But they also were spurious, or wrongly interpreted, or merely coincidences, or approximations, or speculations, or hypotheses, or just manipulated mathematical equations.

Einstein's reasoning was that all of the laws of physics (in the context of velocity) were dependent upon relative velocities and therefore he must apply the so-called 'Lorentz transformations' to such laws.

And sure enough, when Einstein applied the Lorentz transformations to the other perfectly valid laws of physics, they became distorted into invalid laws. He then attempted to mathematically demonstrate that the invalid laws of physics which he had just created were now really valid laws. Upon completing his Special Relativity, Einstein had mathematically turned most of physics on its head, and then presented a wholly spurious theory to the world at large.

The mathematician Bertrand Russell once said that *"Technically, the whole of the special theory is contained in the Lorentz transformations"* (Bertrand Russell, 1927). This is so. In using the Lorentz transformations as the basis for SR Einstein ensured that his SR would be wholly spurious. The

Lorentz transformations are spurious for the following two main reasons:

1. The Lorentz transformations are based on a stationary ether, a concept totally dismissed by contemporary science. This begs the question: if the mathematics of the Lorentz transformations was derived from light movement in a medium (an ether), how can such mathematics directly apply to Einstein's light movement in empty space? Remember that in special relativity an ether does not exist.

2. The Lorentz transformations include concepts of 'absolute true time' and 'absolute local time', which are ad hoc, arbitrary and meaningless concepts.

Yet in 1905 Einstein would adopt the Lorentz transformations as the mathematical foundation for his special theory of relativity and its many bizarre mathematical consequences.

He did this because Einstein could not openly admit the fundamental difference between the two types of light speed, namely 'direct light speed' and 'relative light speed'. For Einstein there was no mathematical difference between both types of light speed.

But why did Einstein think this way? Why exactly was Einstein so mistaken with his SR? The answer is that he made a mistake. Einstein's mistake was that he described and measured the velocity of a light ray with respect to the sun and its motion, instead of with respect to a vacuum (as postulated by James Maxwell). This mistake was Einstein's fundamental false premise for his entire special theory of relativity.

Forty years before Einstein wrote his SR, Scottish scientist James Clerk Maxwell was able to deduce from light experiments that the velocity of a light ray was about 300 million metres per second when it transmitted through a vacuum (denoted as speed 'c' in physics). Maxwell's Law of light speed stands true today.

For several reasons, Einstein did not understand Maxwell's theory for the constant transmission velocity of a light ray with respect to a vacuum. Instead Einstein assumed that the velocity of a transmitting light ray must be described and measured with respect to a material body (such as a rocket), which body might be moving toward or away from the light ray. As a consequence Einstein went down a false path in devising his special relativity.

In effect, what Einstein did was to mathematically change Maxwell's Law so that every light ray in the Universe had absolutely the same constant velocity c (about 300 million metres per second) with respect to every uniformly moving body in the Universe at every instant in time; regardless of the actual different relative velocities of such bodies and regardless of whether or not they were travelling ***toward or away from such light ray***s. This absurd mathematical result is, of course, a physical and mathematical impossibility.

As things stand today in physics there are no empirical confirmations for any of Einstein's distorted concepts or for his Special Relativity. Quite simply, his SR is just an elaborately contrived, spurious and meaningless mathematical theory which attempts to justify its own false premise.

The renowned scientist Nikola Tesla completely rejected the theory of relativity. He insisted that mass and energy were not equivalent and told the New York Times in 1935 that *"Einstein's relativity work is a magnificent mathematical garb which fascinates, dazzles and makes people blind to the underlying errors"* (source: Alex Knapp, 'No, Tesla Did Not Predict Faster Than Light Neutrinos', Forbes.com).

Note: Many scientists in the past and today have expressed their scepticism regarding relativity, including Nobel laureates such as Philipp Lenard, Johannes Stark, Ernst Gehrcke, Stjepan Mohorovičić, Rudolf Tomaschek, Ernest Rutherford, Nikola Tesla and others.

You simply cannot escape the big contradiction between special and general relativity. Special relativity absolutely requires the *non-existence of an ether* because it postulates that the speed of light is constant in a vacuum, i.e. in a space with no ether. On the other hand general relativity absolutely requires the *existence of an ether* because it postulates that objects gravitate towards each other in the presence of each other's mass, and for that to happen there must be an ether. Otherwise, the mass of such objects would not detect each other's presence.

This contradiction has been recognized by scientists and to validate general relativity some people have resorted to 'dark matter', a contemporary equivalent to an ether. As discussed in this book dark matter is completely unproven and has never been detected.

But even if dark matter exists it would need to be everywhere because gravity is everywhere. Wouldn't this invalidate special relativity? How can the speed of light be constant in a vacuum if a vacuum does not exist due to an all-pervading ether in the form of dark matter?

Relativists get round this by saying that the particles of dark matter would be so tiny that they would not affect light, so the properties of a vacuum in space would still apply. But if dark matter particles are so tiny that their mass is virtually unnoticeable, how is it that they can have enough mass to hold galaxies together and give the Universe enough mass to account for gravity everywhere?

There are many intelligent people who today are just repeating and parroting incorrect statements about relativity, such as saying that it has been well tested, when in reality it has never been tested to actually confirm the theory of relativity.

It truly is surprising that the spurious nature of relativity continues to be taught in classrooms as scientific gospel. Like the story of *'The Emperor's New Suit'*, it takes great courage to say aloud *"The Emperor is Naked!"*, particularly when credibility, career and academic sponsorship may be at stake.

This begs the question: *is Special Relativity used in our daily lives? Is SR used in practice in any way?* Are there any instances of SR we can see in our daily lives, and do any technologies we use today demonstrate the veracity of SR? Let's look at some examples put forward by those proclaiming relativity.

Is special relativity used in daily life?

To answer this question the following topics are discussed:

- **A. Electromagnets or magnetism.**
- **B. Speed of light.**
- **C. Sunlight.**
- **D. Solar power.**
- **E. Colour of gold.**
- **F. Laser devises.**
- **G. The physics of financial markets.**

A. Electromagnets or magnetism. We all experience magnetism in one form or another in our daily lives. It is claimed that magnetism is based on relativity. This is a misconception. A magnetic field is not just an electric field with relativity applied. In reality, a magnetic field is a fundamental field which can exist in a certain reference frame without needing any help from an electric field or from relativity. Electric fields and magnetic fields are both fundamental, both are real, and both are part of one unified entity: the electromagnetic field.

Physics textbooks are clear on the matter: magnetic fields can exist without electric fields and electric fields can also exist without magnetic fields. Both magnetic fields and electric fields are not dependent on each other. Anybody who studies electromagnetism will know this.

However, special relativity disagrees: electricity and magnetism are fundamentally interlinked; one cannot exist without the other. What are we to believe?

B. Speed of light. We all experience light every day. It is claimed that relativity makes it possible for light to move at a finite speed. That without relativity, light would move instantaneously, which it clearly does not. The scientific consensus is that photons that make up light do not interact with the Higgs field (the Higgs field gives mass to subatomic particles). As a result, photons do not have enough, if any, proper mass for this to affect their velocity. This means they can only travel at one maximum speed, i.e. the constant speed of light 'c' in a vacuum. Electrons, protons and neutrons all have more mass than photons, therefore can't move as quickly as the speed of light.

Why does light have a finite speed you might ask? The author of this book proposes the possibility that the maximum speed of light is set by the fact that photons do indeed have a tiny amount of 'proper mass' (also referred to as 'intrinsic mass' in physics) but too little to be detected by current science. This tiny amount of proper mass (whether imposed or not by the Higgs field) is what sets the maximum speed of light.

Note: the 'finite' speed of light is the maximum speed of light. The 'constant' speed of light is the unchanging speed of light. This constant speed is set by the atoms that create light (for a more detailed explanation see the section 'The Majesty of Light').

C. Sunlight. The daily sunlight in our lives makes life possible everywhere. It is claimed that without Einstein's most famous equation $E=MC^2$ the sun and the rest of the stars wouldn't shine. In the centre of the sun intense temperatures and pressures constantly squeeze four separate hydrogen atoms into a single helium atom. The mass of a single helium atom is just slightly less than that of four hydrogen atoms. What happens to the extra mass? It gets directly converted into energy, which shows up as sunlight on our planet. The formula $E=MC^2$ is said to explain how this occurs.

In reality, solar energy is created by nuclear fusion that takes place in the sun. Fusion occurs when protons of hydrogen atoms violently collide in the sun's core and fuse to create a helium atom. This process, known as a PP (proton-proton) chain reaction, emits an enormous amount of energy. Einstein's formula $E=MC^2$ is not required for this to happen.

In fact, Einstein's equation $E=MC^2$ (the equation of mass/energy equivalence) has become increasingly suspect. It is thought to be wrong for three main reasons.

"Einstein made three basic mistakes in his interpretation of the $E=MC^2$ equation. Einstein's first mistake with E=MC2 was to take a simple equation and then try to interpret it with two contradictory and paradoxical ideas of mass and energy. Einstein's second mistake with his equation was his failure to realise that the primary meaning of $E=MC^2$ is that it defines the mass of the photon as the truest measure of mass. Einstein's third mistake was to conclude from the Doppler

effect that the motion of light was intrinsically relative and not just hidden from the view of observers. He failed to believe that all photons move at the same speed in a vacuum." Source: James Carter, The Living Universe.

It is now known that mass is a form of energy, but the energy of an object is only ***partially*** determined by its mass, not *entirely* determined by its mass as postulated in the equation $E = MC^2$. That is its main failing.

D. Solar power. In March 1958, the U.S. Navy launched a grapefruit-size sphere dubbed 'Vanguard 1' into orbit around Earth. It was the first satellite to be powered by electric solar cells—shiny slabs of semiconductor that turned sunlight into electricity.

Today, solar cells power almost all the hundreds of satellites orbiting Earth, along with many of the probes being sent to planets as distant as Jupiter. On the ground, solar cells are spreading across suburban rooftops, as rapidly falling prices bring them closer to being competitive with conventional electric power.

It is claimed by some that Einstein invented solar power as evidenced by his Nobel prize. In fact, he was given a Nobel prize for his services to theoretical physics in general, and especially for his explanation of the photoelectric effect. He explained correctly that light itself was just a swarm of discrete energy packets—particles of light that would later be named photons.

Einstein did not invent solar cells; the first crude versions of them date back to 1839, when Edmond Becquerel discovered the principle behind solar energy. A few decades later solar power patents began to be registered by different people in various parts of the world.

Einstein was never given a 'Nobel Prize' for his theories of relativity, and solar power in all its forms does not involve any kind of relativity.

E. Colour of gold. We are all familiar with the colour of gold. It is claimed that the colour of gold is due to relativistic effects. It is claimed by relativists that if you were to calculate the frequency (colour) of gold without taking relativity into consideration, you would perceive it to have a silver sheen. However, the familiar golden colour actually leans further to the red end of the spectrum. It is claimed that the high speed of electrons (in the atoms of gold) makes the electrons contract due to a relativistic effect, and that this in turn changes the colour spectrum to produce the golden colour that we see.

But there is no need to call on relativity to explain the colour of gold:

"The human eye sees electromagnetic radiation with a wavelength near 600 nm as yellow. Gold appears yellow because it absorbs blue light more than it absorbs other visible wavelengths of light; the reflected [incident] *light reaching the eye is therefore lacking in blue compared with the incident light. Since yellow is complementary to blue, this makes a piece of gold under white light appear yellow to human eyes"* (source: Wikipedia.org).

"Silver, gold and copper have similar electron configurations, but we perceive them as having quite distinct colours, depending on how such metals absorb light. Gold has an intense absorption of light with energy of 2.3 eV (from the 3d band to above the Fermi level), so we see it as bright yellow, the colour of gold. But silver for example, has good reflectivity (less absorption) so it appears very close to white" (source: What causes the colours of metals like gold?

webexhibits.org).

F. Laser devices. Many aspects of our daily lives include laser devices of one sort or another, such as fibre-optic cables, laser printers, bar-code scanners, laser speed guns, presentation pointers, laser surgery, and many industrial processes.

Wikipedia describes a laser device as a device that emits light through a process of optical amplification based on the stimulated emission of electromagnetic radiation. Lasers are distinguished from other light sources by expressing their light output as a narrow beam.

Put simply, laser technology is based on being able to make the electrons in atoms all emit the same kind of energy in the same direction in a narrow beam. We won't go into the full details of how laser devices work, but we will comment on the subject in the context of relativity.

Einstein proposed in 1905 that we regard Planck's formula for energy as a physical fact, not just a mathematical trick that avoids the complications of a classical energy continuum. He said that the electromagnetic oscillations of light could be 'chopped up' into a ray of discrete energy packets called photons. This sparked the notion of a laser ray.

From this, relativists claim that the use of lasers is made possible by incorporating the principles of special relativity into laser technology. This is not so. Laser technology does not employ any kind of relativity. No doubt Einstein may have contributed to the eventual development of laser technology, but that is all.

In fact, laser technology is straightforward and a laser is created when electrons in the atoms in optical materials like glass, crystal, or gas absorb the energy from an electrical current or a light. That extra energy 'excites' the electrons enough to move from a lower-energy orbit to a higher-energy orbit around the atom's nucleus.

Relativists make reference to *'relativistic electron beams that provide streams of electrons moving at relativistic speeds'* so as to provide the laser beam.

In particle physics, a relativistic particle is a particle that moves at a relativistic speed; that is, a speed comparable to the speed of light. The phrases 'relativistic particle' or 'relativistic speed' does not refer to or imply any kind of Einsteinian special relativity, it simply refers to particles moving at near the speed of light.

It is well-known that the first laser was built in 1960 by Theodore Maiman at Hughes Research Laboratories, based on theoretical work by Charles H. Townes and Arthur Leonard Schawlow.

G. The physics of financial markets. Albert Einstein developed the fluctuation-dissipation theorem to explain the random movement that was observed in particles found in liquids or gas. This movement was named 'Brownian Motion' in honour of the Scottish biologist Robert Brown who was the first to observe it. It was claimed that the Brownian Motion is not very different from price fluctuations seen in stock markets.

The Brownian Motion basically states that a liquid becomes less viscous (less thick) the smaller the particles. This subject has nothing to do with relativity, but it is mentioned because some relativists are quick to mention the Brownian Motion as evidence of special relativity at work.

The Brownian Motion is not considered suitable for modelling future stock prices because of its independent increments property. This conflicts with how stock prices behave: knowing the present stock price does not necessarily tell you its future price, as implied by Brownian Motion.

No doubt there are other examples that claim to verify relativity in everyday life. But when you drill down into such examples, they quickly prove to be baseless.

For the sake of completeness and to finish briefly on the subject of unravelling relativity, there are three well-known experiments that are cited by relativists to validate relativity: The Michelson–Morley experiment, The Kennedy–Thorndike experiment, and The Ives–Stilwell experiment.

The purpose of the aforementioned three experiments was to measure the velocity of the Earth relative to a luminiferous ether, the presumed medium through which light and electromagnetic radiation are propagated. These experiments are now widely disregarded because they depended on the existence of an ether which is now known to not exist.

There is no attempt to disparage Albert Einstein personally and although his theories of relativity are increasingly viewed as being faulty, he has nevertheless inspired people to take an interest in science.

The reality is that most scientific discoveries and theories are made on the back of existing human endeavour. Analysis of millions of studies and patents has found that nearly all scientific breakthroughs are borne from previous generations of scientists, a pattern that was *'nearly universal in all branches of science and technology'* according to Christian Science Monitor, 2017.

But it doesn't follow that such 'breakthroughs' are necessarily correct or true. In the case of relativity it is well-known that Einstein was influenced by Gauss, Riemann, Mach, Poincare,

Lorentz, Talmey, Newton, Maxwell and others. The fact that Einstein was influenced by such people does not necessarily validate his work. In fact, as discussed in this book, the contradictions and facts regarding relativity are such that it is no exaggeration to say that both special and general relativity are wholly spurious, not just needing to be tweaked, updated or re-interpreted.

Many scientists today are ensconced in research institutions that must follow the so-called 'standard model' of scientific-thinking for fear of losing research grants or not being a good 'team player'. Scientists looking for advancement have to tread carefully for fear of being thought a crackpot if going too far outside accepted dogma, something better not to touch if a career and credibility is desired. They need a post-doctoral or better position. To ensure success they do what is needed. This partly explains why baseless Einsteinian relativity continues to persist in scientific circles today.

The many so-called experimental confirmations of the special theory of relativity have greatly helped to turn Doubting Thomases into faithful, unquestioning relativity believers and Einsteinian advocates. But when fully scrutinised, it turns out that all of these so-called confirmations are in reality coincidental approximations, rank conjectures, empirically invalid concepts, misconceptions, illogical interpretations, ad-hoc mathematical consequences, circular reasoning, explainable paradoxes, untestable speculations, wishful thinking, and the like.

The takeaway message: both special and general relativity are shown to be spurious theories for a variety of reasons. We should no longer torture physics students with incomprehensible fairy tales about relativity. The study of astrophysics will continue to be held back for as long as it remains tethered to Einsteinian relativity.

The Accordion Theory of the Big Bang

The accordion theory of the Big Bang revealed here discusses how the Universe will end and what came before the Big Bang. Clearly, any theory on this topic is going to be entirely speculative, mainly because of the timescales involved. Nevertheless, by taking into account contemporary cosmology theories on the subject and combining this with the *Final Theory of Everything* a most likely scenario can be put forward which is referred to as the accordion theory.

Cosmologists have wrestled for many years with the dual question: how is the Universe likely to end and what came before the Big Bang? To answer this question we will first look at how the Universe will end, and then we will consider what came before the big bang. As explained in the accordion theory, both these events are intimately linked as one follows the other.

The fact that the Universe is expanding at an accelerated rate means that all the matter in the Universe becomes spread out over more and more volume with each passing day.

Matter and energy can change, but it is not thought that matter created from nothing can come about or that matter can completely cease to exist. So although new stars are always being created in the Universe such stars are created from existing materials. Put more formally, we live in a null-energy Universe in which the total energy of the Universe stays balanced at zero even though matter and energy in all its forms can be changing in various ways.

We currently live in what cosmologists call the luminiferous era — the epoch of our Universe that is full of stars, light and warmth, so is the Universe past its prime? It is thought that star formation peaked almost ten billion years ago and has been declining ever since. The reason for this strange dwindling of light is the fact that we live in an expanding Universe, but the amount of matter within the Universe remains fixed. The AE force postulated in this book makes everything that exists become more and more dispersed.

But it may seem that the opposite is happening. That stars and galaxies are coalescing. For example, it is thought that our local group of galaxies, consisting of the Milky Way, Andromeda, and the Triangulum galaxies, along with dozens of satellite dwarf galaxies, will remain nearby or gradually come together. But even these 'local' galaxies will eventually stop coalescing and thin out.

New star formation is dwindling because to make a star, you need to compress matter down into relatively small volumes, but as the Universe ages there are simply fewer and fewer opportunities for this to happen.

It's difficult to speak of the very far future of the Universe with any level of precision, but we can make rough estimates and the FTOE helps us to do this. Our cosmos is currently thought to be nearly 13.8 billion years old, and galaxies throughout the Universe will continue making new stars for many years to come. But eventually — very roughly one trillion years from now — the last star will be born.

That star will likely be a small red dwarf, barely a fraction of our sun's mass. Red dwarf stars live fantastically long lives, gently sipping on hydrogen to power a slow but steady fusion reaction. But eventually, all stars, including the red dwarfs, will come to an end. In roughly 100 trillion years the last light will go out.

As the luminiferous era slowly unwinds, the Universe itself will change in character. The current size of the observable cosmos, defined in part by the most distant objects that we can see, is roughly 90 billion light years across. The volume bounded by that diameter contains about two trillion galaxies. So the following two questions arise:

1. Will all the matter in the Universe become more dispersed to the extent that new star formation will cease?

2. And what will eventually happen to this dispersed matter?

The FTOE postulated in this book can help us speculate as to what is most likely to occur (given our current knowledge) and answer these two aforementioned questions.

The AE force (Accelerating Expansion force of the Universe) is likely to continue for many years but we have no idea for how long. Eventually, the initial impetus of the Big Bang that galvanised the accelerating expansion of the Universe may very gradually fade away to a point where the Universe is no longer expanding.

Without such accelerated expansion gravity as we know it will disappear in all parts of the Universe. And without gravity galaxies and orbiting objects will disintegrate even at a subatomic level. All matter in the Universe will 'thin out' and all biological life will cease to exist everywhere.

To continue with our speculation, as cosmic expansion comes to an end the Universe will gradually enter a phase of contraction. The change from expansion to contraction will be very gradual, over a very long time. So we have four main phases in what we are calling the 'Accordion Theory of the Big Bang'. These phases are as follows:

Phase One: Rate of expansion slows down.
Phase Two: Cosmic expansion ceases.
Phase Three: Cosmic contraction.

Phase Four: A new Big Bang.

Note: The *'Accordion Theory'* is so named to differentiate it from the 'Cyclic Universe Theory'. Both theories agree that the Universe undergoes endless cycles of expansion and cooling, each beginning with a new big bang and ending in a big crunch. But in the accordion theory stars and planets are reborn as the Universe contracts over billions of years heading towards the big crunch. A whole new Universe emerges with galaxies and life similar to the Universe that we see today.

The accordion analogy arises from a musical accordion that plays melodies whether the accordion bellows are expanding or contracting.

Phase One: Rate of expansion slows down. The current rate of acceleration (of cosmic expansion) will gradually slow down over many millions of years until it is no longer accelerating. This will happen because the original impetus of the Big Bang that galvanises the accelerating expansion will gradually fade away. This phase may already have started but is not noticeable on a human time scale (or this phase is yet to come).

As cosmic expansion slows down, the force of gravity will also diminish everywhere. Remember that the FTOE tells us that cosmic expansion and gravity go hand in hand. Without the AE force there can be no gravity even at a subatomic level. However the rate at which cosmic expansion slows down over many millions of years would not be noticeable by any intelligent race thriving at the time.

Phase Two: Cosmic expansion ceases. When the Universe is no longer expanding at an accelerated rate it will gradually begin to contract (phase three). The Universe will not simply stop expanding and then stay at the same size forever. Why not? Here is the answer:

1. As cosmic expansion ceases, the matter in the Universe will continue to move, i.e. meander 'aimlessly' in the absence of the AE force. Such movement continues because of legacy momentum. A scenario in which all matter in the Universe becomes frozen in time (i.e. not moving) is very unlikely to occur. The continued movement of matter acts to prevent a Universe that stays forever 'frozen in amber'.

The original outward movement of all matter in all directions caused by the Big Bang is what enables the Universe to expand into new space. This movement carries momentum which continues even when cosmic expansion ceases. Such movement (arising from momentum) will very gradually lead to cosmic contraction arising from the inertia/gravity of such movement. Hence, the matter in the Universe does not stay 'frozen' and 'immobile' forever. Such matter (objects) will continue to move because of their legacy momentum.

2. If it is postulated that the Universe will end up staying the same volumetric size, it doesn't explain how the Universe started, i.e. what caused the Big Bang? It implies a negation of the Big Bang. However, if it is postulated that eventually the Universe will contract into another Big Bang, then we have an explanation as to how the Universe started. We can postulate that in accordance with scientific observations, a Big Bang created our Universe about 14 billion light years ago, and that the Big Bang was caused by the contraction of a previous Universe. In scientific circles this is known as a 'cyclical Universe' and either this or the 'accordion theory of the Big Bang' is the most likely scenario.

3. There is an argument saying that the Universe can indeed end up staying the same size because the 'self-gravity' of all matter is sufficient to keep it that way (the so-called 'critical

density theory'). This theory assumes the existence of a 'self-gravity' phenomenon that depends on space curvature as the cause of such gravity.

"Self-gravity is the process by which objects are held together by the combined gravity of such objects as a whole. Without it, stars, stellar clusters, galaxies, and groups and clusters of galaxies would all expand and dissipate" (source: Critical Density, the SAO Encyclopedia of Astronomy).

To accept self-gravity as an explanation for a static Universe you would have to accept the concept of gravity based on spacetime curvature and the concept that the Universe will stay the same size forever. It is speculated that this is not so.

Another theory currently doing the rounds in cosmological circles says that everything that exists will eventually evaporate, leaving nothing behind. Here is an abridged extract from a study on this subject:

"Black holes emit a type of energy called Hawking radiation. This depletes a black hole, causing it to gradually evaporate. It is suggested that over trillions of years the whole Universe will just evaporate into nothing. That particles in everything can be ripped apart just solely from gravitational forces in the vacuum" (source: Michael F. Wondrak, et al, Gravitational Pair Production and Black Hole Evaporation, Physical Review Letters 130, 221502, 2023).

This Wondrak study does not make it clear whether all matter in the Universe will completely cease to exist in any form, or whether all matter will end up as some kind of radiation/energy. Other researchers disagree with the study saying that a more careful analysis will show that the particle-antiparticle pairs discussed in the study don't actually radiate from massive non-black hole objects. It should also be mentioned that the study calls upon Einsteinian spacetime to validate its conclusions.

Phase Three: Cosmic contraction.

It is thought by some scientists that the ultimate fate of the Universe depends on the amount of mass contained within it, that is, the mean density of the Universe. There is a critical density, which if reached will be sufficient to just about stop the expansion of the Universe.

The 'critical density' is said to be the average density of matter required for the Universe to just about halt its expansion, but only after an infinite time. A Universe with critical density is said to be flat.

However, it is speculated that the ultimate fate of the Universe does not depend on the amount of mass created within it. Otherwise you have to accept that gravity is caused by a curvature of space (the general theory of relativity), and that cosmic expansion will cease after an 'infinite' amount of time. To say that cosmic expansion will only cease after an infinite amount of time is the same as saying that cosmic expansion will never cease, clearly absurd.

We therefore discount theories that talk about 'critical density' or that depend on General Relativity and the curvature of space.

We speculate that although UI (Universal Inertial) Gravity disappears as cosmic expansion disappears, a small amount of gravity remains in all the moving matter in the Universe. There is no evidence or reason to think that all matter in the Universe will eventually come to a halt (stop moving). Such matter will always be moving as a result of their legacy momentum.

Note: In current cosmology it is accepted that a moving object in space that has no means of self-propulsion will continue moving indefinitely if it does not encounter other objects, energies or effects of gravity.

It is well-established in physics that momentum and inertia are related because the more momentum an object has, the more inertia it has as well. Both properties are dependent on the mass of the object. The more mass in an object, the higher its momentum and inertia.

The mentioned 'legacy' momentum is speculated to be enough to enable such matter to gradually coalesce. As the original momentum of the Big Bang is no longer there, the outward movement of matter that required cosmic expansion no longer applies. Consequently, the gradual coalescing of 'meandering' matter arising from legacy momentum will result in such matter coalescing. This in turn will shrink the total volumetric size of the Universe that is no longer expanding, and very gradually cosmic contraction will ensue.

When the Big Bang occurred there was no empty space already in existence waiting to receive the Big Bang. The outward movement of super-hot plasma, which occurred as a growing circle on a plain, is what triggered the cosmic expansion that we see today. Equally, as the Universe contracts in this phase three, it will not leave behind a Universe of empty space. The whole Universe will become smaller as it contracts.

The rate of contraction will accelerate at a constant rate just as the rate of expansion accelerated at a constant rate. This will in turn cause a renewed inertia and gravity in all parts of the contracting Universe.

There is some evidence for this assertion. The Friedmann–Lemaître–Robertson–Walker metric on which the standard model of cosmology is based describes a homogeneous expanding Universe with a metric that can equally apply to a contracting Universe.

Why would such contraction be at an accelerated rate? Because mathematically, with each passing moment the total percentage of contraction must go down more and more. Let's suppose the contraction starts at moment 10. From moment 10 to moment 9 the total percentage-contraction of the whole Universe is say, 2%. So the Universe has contracted from size 100 to size 98. From moment 9 to moment 8 it contracts another 2% (but that is 2% of 98, not 2% of 100). So from moment 9 to moment 8 the contraction is 2% of 98, which is 1.96 in size, and 1.96 is less than 2.

So even if cosmic contraction is the same 2% with every passing moment, that 2% will equate to a smaller and smaller amount of total volume, i.e. an accelerating reduction in volume.

As cosmic contraction ensues we can further speculate that it will give birth to new stars, galaxies and biological life during the contraction phase. New stars and planets could come about because cosmic contraction will again create UI gravity. It will be a kind of cosmic expansion in reverse because the rate of cosmic contraction will be at an accelerated rate. And this accelerated rate provides the same kind of gravity and conditions for life as under accelerated expansion.

Thus, this accelerated rate of contraction over many billions of years could give birth to a new 'Universe' similar to the Universe of today. This is why we are dubbing it the 'accordion theory'.

Phase Four: A new Big Bang. Eventually, over many billions of years, the contraction of the Universe will reach a point in which the Universe becomes too small to sustain life or the creation of stars and planets. This brings us to the big crunch.

"The Big Crunch is a hypothetical scenario for the ultimate fate of the Universe, in which the expansion of the Universe eventually reverses and the Universe recollapses, ultimately causing the cosmic scale factor to reach zero, an event potentially followed by a reformation of the Universe starting with another Big Bang" (source: The Big Crunch, Wikipedia.org).

There is some research claiming that the Big Crunch Theory is unlikely because so-called astronomical observations show that the expansion of the Universe is accelerating rather than being slowed by gravity, suggesting that the Universe is far more likely to end in heat death, i.e. a death caused by lack of heat everywhere.

But this is faulty reasoning. Cosmic expansion will slow down eventually, but not because of cosmic gravity or lack of it. There is no evidence pointing to some kind of omnipotent cosmic gravity emanating from the sum of all gravities in the Universe.

Thus, cosmic expansion will not slow down from the fading away of some mysterious all-powerful force of gravity. It will slow down because the original impetus of the Big Bang that galvanises cosmic expansion will fade away.

It is understandable that when astronomers see an expanding Universe, there is no reason to suppose it will ever stop, so most astronomical observations conclude incorrectly that this will continue, that matter will become more and more dispersed and 'cold' and hence the name 'heat death' or 'big freeze'.

In this phase four, as the Universe contracts, becoming smaller and smaller, a point will be reached that creates the conditions for another Big Bang. Matter will become so dense that it can no longer contract. The result will be another Big

Bang, but not like a big supernova explosion. It will be more akin to a very, very super-hot point springing into being and then rapidly growing in all directions on a plane (spreading out like a pancake rather than like a sphere). This is how it is thought that our current Universe started in the so-called big bang.

What came before the Big Bang?

Something caused the Big Bang, but what? The answer that follows is considered to be the most likely explanation in contemporary cosmology. Nevertheless, we are still in the realms of speculation.

The standard model of the Big Bang and the consensus among most cosmologists is that the Big Bang was not an enormous explosion, with debris flying out in all directions to later become galaxies. This scenario is very unlikely because it would require the prior existence of empty space in which such an explosion could occur. When the Big Bang occurred there was no empty space waiting to receive the Big Bang.

The mentioned accordion theory proposes a much more likely scenario that fits in with the standard model of cosmology on this subject: the appearance of a super-hot point that appeared in the midst of nothingness. When this hot point appeared it made its own space in which it could appear, because when it appeared there was nothing, not even empty space. This very, very tiny hot point was endowed with the same Accelerating Expansion force (AE force) that we find in our present-day Universe. The 'inbuilt' AE force of the tiny super-hot point gave it the ability to create new space in which it could exist and grow.

Our Universe is becoming bigger and bigger with every passing moment as evidenced by the redshift, the Hubble

Constant and other well-studied phenomena in cosmology. The Universe is compelled to make itself bigger arising from the original heat-energy impetus of the Big Bang. Think of bread-dough that grows in size from the heat it receives. But there is nothing beyond the overall size of the Universe, not even empty space. So as the Universe grows in overall size it has to create more empty space into which it can grow in size.

We humans do not understand (yet?) the mechanism by which the Universe creates more empty space from nothingness, but all our cosmological observations show that this is indeed what is happening. The most likely explanation is that new space is created as a result of negative pressure.

In physics and engineering a negative pressure arises when an enclosed volume has lower pressure than its surroundings. Thus, it is speculated that negative pressure arises when the whole volume of the Universe has a lower pressure than the surrounding 'nothingness'. To relieve this negative pressure the Universe 'creates' or 'takes' space from the surrounding nothingness. This of course is pure speculation, but it is the best explanation we have in contemporary cosmology. This topic is examined in great depth in the excellent book 'From Here to Infinity' by Bruce Jimerson.

But what came before the mentioned super-hot point that appeared into nothingness to eventually form our Universe? What exactly caused the super-hot point to appear to begin with? Here is the answer.

As mentioned, it is speculated that the Universe will end as super-dense, super-hot energy that will be contracted into a very tiny point. Let's call this an 'accordion-point'. Outside of this accordion-point there will be nothing, not even empty space. This mentioned accordion-point of very, very dense hot energy will not be a so-called singularity as defined by relativity. It will not be a point of infinite density and infinite gravity within which no object inside can ever escape, not even light.

The accordion-point will not be constrained by any kind of strong gravity because there will be no gravity in this scenario. You will know that gravity is caused by inertia, which in turn is caused by accelerating movement as explained in this book. But the accordion-point is not moving because there is no empty space in which to move, and hence there is no force of gravity preventing it from springing back so to speak.

It is speculated that the accordion-point will be so hot and so dense that it will become endowed with unknown laws of physics such as the AE force. More specifically, the accordion-point will reach a moment when it cannot go on becoming hotter and denser, so it will 'spring back' to form a new and growing point of energy in a place of nothingness. This re-surfacing or 'springing back' of the accordion-point is what we call the Big Bang.

This 'springing back' phenomenon is well-known to science because this is how all stars are born. As a gas cloud becomes denser and hotter (caused by gravity) it will reach a point when it can no longer become denser and hotter. At this point the very dense and hot energy will 'spring back' to form a shining star. This springing back is caused by nuclear fusion. In the case of the Big Bang, the springing back of the accordion-point will also be caused by nuclear fusion.

Many studies speculate that the Big Bang was galvanised by nuclear fusion, and this is the consensus opinion in contemporary cosmology. But although both the Big Bang and all the stars in our Universe are said to have come into being from nuclear fusion, the nuclear fusion of the Big Bang was a little different by virtue of being on such a bigger, more profound scale. In astrophysics the nuclear fusion of the Big Bang is referred to as 'Big Bang nucleosynthesis' or BBN.

This new accordion-point of the Big Bang will be endowed with the AE force, giving it the ability to create new space in the middle of nothingness, thus giving the accordion-point room in which to exist and grow. So what came before the Big Bang? What came before was the previous Universe that had contracted to a very tiny super-hot, super dense point of energy. This same accordion-point then sprung back as a new Big Bang to form a new Universe.

In short, when the Universe ended, there was never a moment of complete nothingness in which even empty space did not exist. The accordion-point of the old Universe never completely ceased to exist; it was reborn as a new Universe. This is the answer to the question: *what came before the Big Bang?*

So from the old Universe that we inhabit, a new Universe will be born and the cycle is repeated. It is speculated that this is how our 'accordion Universe' will end and be re-born.

We will never know whether the Universe that we inhabit is the first, the tenth, or the trillionth Universe in this cycle of repeated Universes. And as to the question of what started off the cycle of Universes to begin with - this may be best left to philosophers.

***

Please see the next page for a short thank you message from the author ➞

Message From Author

Thank you for reading the *Final Theory Of Everything*. If you liked the book please leave a brief review at the online or offline place of your book purchase. Also, any feedback, positive or negative, is much appreciated as it will help improve and update future editions - for this, my email is: mailto@deliveredonline.com (please put only the title of the book in the email subject heading to make sure I get it).

Here is the publisher's website: www.DeliveredOnline.com in case you would like to tell others about this book.

Russell Eaton, author

***

Author Bio

Russell Eaton is British and the author of several non-fiction books, mostly relating to health and well-being. With a passionate interest in cosmology, Eaton's book 'Final Theory of Everything' is his biggest project.

He has lived in the UK and in Ecuador, splitting his time between the two countries, and sometimes getting into a pickle because of it. Widely travelled, Russell Eaton keeps an interest in any and all the wonders of the world and the Universe.

He says "We must always endeavour to eradicate bigotry and prejudice from science, and we must always be on guard when such pernicious influences come knocking at the door".

***

www.ingramcontent.com/pod-product-compliance
Lightning Source LLC
Chambersburg PA
CBHW071534200326
41519CB00021BB/6489

* 9 7 8 1 9 0 3 3 3 9 7 6 3 *

9 781903 339763